RENEWALS 458-4574
DATE DUE

WITHDRAWN
UTSA LIBRARIES

JAPAN AND THE ASIAN PACIFIC REGION

JAPAN and the ASIAN PACIFIC REGION
Profile of Change

SHIBUSAWA MASAHIDE

ST MARTIN'S PRESS
New York

© Royal Institute of International Affairs 1984
All rights reserved. For information, write:
St. Martin's Press, Inc., 175 Fifth Avenue, New York, NY 10010
First published in the United States of America in 1984

Library of Congress Cataloging in Publication Data
Main entry under title:
Shibusawa, Masahide, 1925-
   Japan and the Asian Pacific region

  Bibliography: P.
  1. East Asia—Politics and government. 2. Asia,
Southeastern—Politics and government. 3. East Asia—
Foreign relations—Japan. 4. Japan—Foreign relations—
East Asia. 5. Asia, Southeastern— Foreign relations—Japan.
6.Japan—Foreign Relations— Asia, Southeastern
I. Title
DS518.1.S476   1984   327.5205   84-16043

ISBN 0-312-44049-9

Printed and bound in Great Britain

CONTENTS

Acknowledgements

1	**Introduction**	1
2	**Cold War in Asia: The Formative Years**	10
	The Korean War	12
	The Polarization of Japan	15
	The Emergence of Economism	22
	The Vietnam War	26
	Lessons of the War	30
3	**Asia Goes Multipolar**	35
	Change in Indonesia	35
	The Birth of ASEAN	37
	Japan's Growing Involvement in the Region	41
	Détente and Balance-of-Power Politics	46
	Dilemma in the Korean Peninsula	51
	A Zone of Peace, Freedom and Neutrality	57
4	**Japan in the 1970s**	62
	Sino-Japanese Relations	63
	Discord in US-Japanese Relations	68
	Anti-Japanese Movements in Southeast Asia	73
	The Oil Shock	78
5	**A New Paradigm of Relations**	86
	The Bali Summit	88
	The Defence Debate in Japan	92
	Japan's 'Equidistance' Diplomacy	97
	The Fukuda Doctrine	101

6	**Persistent Conflicts and Communism in Asia**	110
	Indochina	110
	The Korean Peninsula	117
	Hong Kong	122
	Communism in Asia	130
7	**The Regional Economy and Interdependence with Japan**	136
	ASEAN	139
	Pacific Economic Cooperation	144
	Japan's Involvement in the Regional Economy	148
8	**Japan's Place in the World**	157
	Isolation	158
	Dependence	162
	Security	165
	Regional Relations	170
	Economism Whither?	176
Notes		179
Bibliography		186
Index		191

ACKNOWLEDGEMENTS

Besides the invaluable support given to me by the staff of Chatham House, I have been fortunate in having a multinational network of friends who kindly read the manuscript at one stage or another: Wilber A. Chaffee, Jr, Derek Davies, Emma Gee, Geoffrey Goodwin, Han Sung-joo, Fred Hendricks, András Hernadi, Ichioka Yuji, Khien Theeravit, Kimoto Tsukasa, Lee Poh Ping, Likit Dherawegin, Robert Longmire, Hans Maull, Wolf Mendl, Charles Morrison, Narongchai Akrasanee, Nishihara Masashi, Okabe Tatsumi, Sarasin Viraphol, Sekiguchi Sueo, Tom Smith, Somsakdi Xuto, Sukhumband Paribatra, Jorgen Thygessen, R. Vaitheswaran, William Wallace, David Watt, and Yamamoto Tadashi. Although they will find the final product to be rather different from what they read, I am greatly indebted to them for their comments and criticisms, which proved indispensable for organizing the themes and thought of the book. Needless to say, the responsibility for errors and omissions rests completely with me.

I am grateful to Gordon Daniels, Ronald Dore, Stuart Harris, Michael Leifer, Ian Nish, Galina Orionova, Satoh Yukio, Radha Sinha, and many others who attended Chatham House seminars and study-group meetings and gave me ideas and insights. Also, I am grateful to Maureen White, who provided me with valuable data and information on the regional economy.

The following officials were kind enough to grant interviews and answer many of my questions: Prime Minister Dr Mahathir of Malaysia; Foreign Minister S. Dhanabalan of Singapore; Ambassador T.T.B. Koh of Singapore and Ambassador Birabhongse Kasemsri of Thailand, both at the United Nations; Kiuchi Akitane, Japanese Ambassador to Malaysia; K. Kesavapany of the Singapore High Commission in London; Muhammad Yusof Hitam and James Gan in the Malaysian Ministry of Foreign Affairs; Hanabusa Masamichi, Nishiyama Takehiko, and Ogura Yoshikazu in the Japanese Ministry of Foreign Affairs.

During the course of the work, I was privileged to participate in, and to draw ideas and information from, the meetings of the Trilateral Commission, the Korea-Japan Intellectual Exchange, the European-Japanese Conference (Hakone VI), the Thai-Japanese Conference, and a series of meetings and workshops of Asia Dialogue, many of which were ably organized by the Japan Center for International Exchange.

This book would not have been possible without the extraordinarily generous contribution of Brian Bridges, whose knowledge of the region, skilful research and, above all, friendship were an incalculable asset thoughout the endeavour. Also, it would not be half as presentable as it is without the superb expertise and attention to detail which were applied to it by Pauline Wickham, who travelled all the way to Tokyo for two months of vigorous editing.

Finally, I would like to dedicate this book to my wife, Chako, who consented to come to live in London for many lengthy periods, and to travel with me frequently around the world, and gave me unceasing support and encouragement.

January 1984　　　　　　　　　　　　Shibusawa Masahide

TO CHAKO

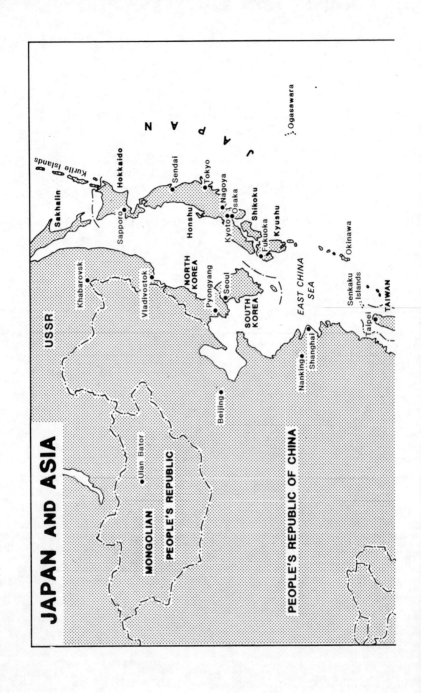

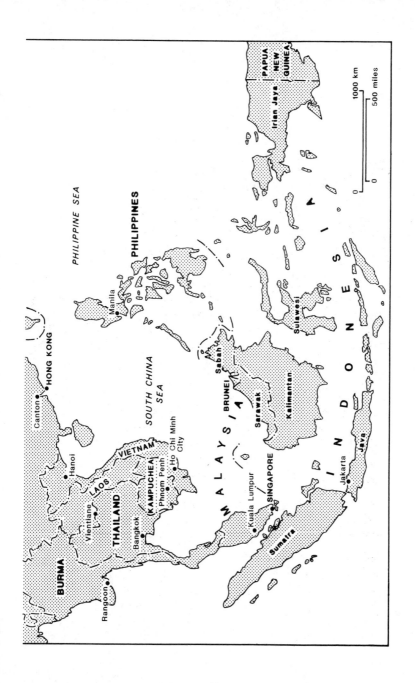

Note: Throughout this book, Japanese names are shown with the family name first and the personal name second.

Chapter One
INTRODUCTION

This book was conceived with two themes, and two aims, in mind. One is to follow the political and economic changes that have occurred in the Asian Pacific region since World War II and to consider their implications; the other is to examine, against this backdrop of change and development, the path trodden by Japan, its position and the problems it has faced, as well as to speculate on its future. Despite its far-flung activities, Japan seems to have particular difficulty in finding its place in the world. By considering how post-war Japan perceived the events occurring in the region and how it responded to them, this book seeks to find clues to an understanding of the basic pattern of its relations, and of its evolving role in the region and in the world.

These two themes are not always complementary. Even though geographically Japan is part of the region, its post-war development has been quite unlike that of the region as a whole. In this respect, it stands in marked contrast to West Germany (with which certain parallels can be drawn), which was conceived from its very foundation as an integral part of the West European system.

Japan belongs to no such integrated system. Nor is there any cultural, economic or political homogeneity in the region. The Asian Pacific countries are very diverse in their culture and history; the process of their post-war development, both economic and political, has been unstructured and erratic. Even the communist countries have shown little solidarity, and, what is more, their relationship with the Soviet Union has been marked by violent fluctuations. Likewise, the non-communist countries have held a variety of political stances, ranging from anti-communist to neutral, so that there was little to encourage them to form a united front.

The term 'Asian Pacific region' is used in this book to refer primarily to the countries which border the western rim of the Pacific Ocean, namely Japan, North and South Korea, China, Taiwan and Hong Kong, through to the Indochinese states and the countries of ASEAN (Association of South-East Asian Nations), namely Indonesia, Malaysia, the Philippines, Singapore and Thailand. From Japan in the north to the southern border of China is referred to as Northeast Asia, and the countries south of Hong Kong are classified as Southeast Asia. China is dealt with mainly in terms of its interaction with, and impact on, Japan and other countries in the region. The activities of the two superpowers, the United States and the Soviet Union, and other countries in and around the region will be examined as and when their activities are relevant to the theme of this study.

* * *

Northeast Asia has a vital importance for the world as the centre of the triangular relationship that exists among the three major powers, the United States, the USSR and China, each of which has extensive interests and commitments in the region. Any change which occurs in the relationship among these three powers is bound to have a far-reaching impact on the global equation of power. The presence of Japan, with its vast, and yet somewhat vulnerable, economic reserves, adds to the complexities of the region.

Ever since the heyday of Western imperialism, Northeast Asia has aroused keen competition among the major powers of the world. The inertia which prevailed in the internal politics of China and Korea in those days delayed their response to the new era of modernization and industrialization with all its dangers and possibilities, and thus critically aggravated the instability of the region. In a desperate effort to expand their respective spheres of influence, Japan and Imperial Russia fought a major war at the turn of the century. As a result of its victory in that war, Japan asserted its dominant position in the Korean peninsula and northeastern part of China, at least temporarily, but it failed miserably in reducing the region's basic instability. The United States, which joined the ranks of the imperialist powers after its acquisition of the Philippines, opted to involve itself in the region with a curiously moralistic policy of upholding the 'territorial integrity' of China against the aggressive designs of other imperialists, notably Japan. After a long and painful succession of events, the region was engulfed by a series of wars which were to last for many years, ravaging the length and breadth of the region, both Northeast and Southeast Asia.

After World War II, with Japan laid low, the United States

assumed that the stability of the region could now be maintained by means of an understanding with the Soviet Union that it would limit its activities in the Korean peninsula to the area north of the 38th parallel. The United States also counted on China to emerge, under the Kuomintang leadership, as a pillar of 'democracy' in the region. Such wishful thinking, however, was rudely shattered by the victory of the communists in China and the subsequent war in Korea, which had the effect of spreading the cold war throughout the world. Alarmed by the prospect of a communist takeover of the entire Asian continent, the United States, under the banner of 'containment of China', launched a crusade which was to have a decisive influence on the life of the region for many years to come.

* * *

Southeast Asia, in contrast with tension-ridden Northeast Asia, is characterized by its 'soft' structure. The ethnic and cultural patterns of the region are marked by their extreme diversity. A population of over 300 million, representing a myriad ethnic backgrounds and historical experiences, embraces all the major religions of the world. National institutions are equally diverse in form and substance, ranging from monarchy to military autocracy to communism. Even the notion of a 'nation', or the concept of a central government, was until recently confined to the elite in capital cities, while the peasants and villagers had little knowledge of, let alone direct dealings with, government institutions. This environment, coupled with easier access from Europe, facilitated the process of colonization, so that, with the exception of Thailand, the entire region was under colonial domination until the end of World War II.

As a result, the post-war history of Southeast Asia evolved either around the struggle for independence or its aftermath, bringing into its midst the communist movements, which fed largely on the fervour of nationalism, or around the aspirations of the masses to change the *status quo*. In North Vietnam, the communists succeeded in establishing their own government, whereas in many other countries they ended up competing for power with the newly born nationalist governments, which consisted, for the most part, of the traditional elite, who had a tendency to preserve their own status and privileges. In these cases, the communists posed a considerable threat to the governments concerned by organizing armed insurgencies or supporting dissident groups, with the covert or overt assistance of communist powers both inside and outside the region. This was interpreted by the United States and its allies as part of a grand communist design to dominate the whole world. Initially, the former colonial powers took up the fight against the communist threat, partly to preserve their own interests, but, following the defeat of the French in Vietnam, the United States was inexorably drawn into the region. Thus the whole of Southeast Asia became embroiled in the cold-war confrontation.

* * *

'When elephants fight, the grass is crushed.' These words, purportedly coined by Lee Kuan Yew, the Singapore Prime Minister, describe the hazards which descend upon neighbouring small countries when superpowers clash. The worst type of hazard is to become a proxy. Being at the centre of a superpower conflict, South Vietnam had to undergo every conceivable suffering, only to disappear from the face of the earth. To take sides with the United States was also a precarious enterprise, since it tended to aggravate internal ideological polarization. Japan experienced violent internal convulsions for precisely this reason during the 1950s. South Korea and Taiwan, which, because of their geopolitical circumstances, had no alternative but to depend on US protection, found themselves having to suppress internal opposition by force. Thailand managed better, but was compelled to walk a political tightrope in order to maintain as much freedom of action as it could, while at the same time cooperating in the US war effort.

One characteristic of superpower behaviour is a penchant for unilateral action. Superpowers differ from regional or small powers in that they formulate their policy on the basis of a perceived threat arising from countervailing superpower actions. The prospective impact of their policies on the countries of the region is only a secondary consideration, because, it goes without saying, global responsibilities and dangers have priority over regional concerns. For instance, the way the United States decided on the 'containment of China' policy was as unilateral as when it later switched to a policy of rapprochement with China. It was a brilliant move on the part of the United States in that it helped to extricate it from the costly war in Vietnam and to break the impasse in global politics. However, the impact on the countries in the region, because the decision was so unexpected, was just as great as its action in Vietnam had been.

The continuation of Lee Kuan Yew's metaphor, 'When elephants make love, the grass is also crushed', aptly describes the aftermath in the region of the Sino-US rapprochement. For example, Taiwan's status suffered badly in the international community and has hardly recovered since. The shock to the two Koreas manifested itself in the internal confusion of both countries during the 1970s. Even North Vietnam was not exempt. The special prestige that it used to enjoy as champion of the struggle against the forces of imperialism was taken away by the single stroke of Nixon's overture to China. Although the subsequent American withdrawal enabled North Vietnam to achieve its long-standing national goal, namely unification with South Vietnam, the country was virtually relegated to the position of a provincial actor suffering from a provincial conflict with its communist neighbours.

ASEAN was formed with a view to minimizing the risk of great-power confrontation by institutionalizing the indigenous framework of cooperation among the non-communist nations of Southeast Asia. Fortunately for ASEAN, in the environment of the mid-1970s, the great powers preferred the neutrality of the region to the alternative case of instability, which would have compelled them to intervene — if only to pre-empt any competitive moves being made by their counterparts. In other words, one of ASEAN's aspirations — to make Southeast Asia a 'zone of peace, freedom and neutrality' — turned out to coincide with the interests of the great powers. Therefore, the ASEAN governments began to concentrate their attention and energy on social and economic development, which was considered to be the logical prerequisite for the region's stability and resilience.

It may well be that, in a sense, ASEAN was able to realize, in a different form, what the non-aligned movement (NAM) was aspiring to in the 1950s. The concept of non-alignment included a genuine aspiration to check the spread of the cold war, and to avoid undesirable superpower intervention. Later, however, when leading countries such as India and Indonesia shifted further and further to the left, the movement was relegated to the level of, at best, idealism and, at worst, a tool of communist forces. In this context, the enviable performance of ASEAN in the second half of the 1970s, shown in its impressive economic growth as well as in its broad acceptance by the international community, seems to have had some impact on countries outside Southeast Asia. Sri Lanka began to sound out the possibility of joining ASEAN, and other neighbouring countries, such as

Fiji, Papua New Guinea and Brunei, sent observers to some of the ASEAN meetings. Also, recent moves among South Asian countries, namely India, Pakistan, Sri Lanka, Bangladesh, Nepal, Bhutan and the Maldives, to form an organization for South Asian regional cooperation (SARC) — more or less similar in concept to ASEAN — indicates the strength of ASEAN's appeal as a model for other developing countries.

* * *

It is ironic that both of the two fiercest ideological wars of the postwar world were fought in the Asian Pacific region despite the fact that communism was a Western idea, which originated in the socio-economic conditions of nineteenth-century industrial society. Also, it is significant that, whereas the communist side chose to fight these wars in Asia via its proxies, namely North Korea and North Vietnam, on the Western side the United States itself joined the wars, thereby inflicting serious damage on its own national strength in the process. However, the communist camp did not emerge unscathed. The succession of events from the Sino-Soviet rift, through to the Khmer Rouge massacres in Kampuchea,* and the three wars in Indochina, led to a serious deterioration in the international prestige of communism. Also, the economic stagnation which perennially plagued all the communist countries in the region shattered the image of communism as a model for economic development. It is indeed doubtful whether communism could ever recover the infectious appeal as a revolutionary proposition which it used to enjoy among the youth and intellectuals of the region.

As though to prove the point, China is engaged in the ambitious and difficult experiment of trying to reconcile, and combine in its national structure, three conflicting factors: the political system of communism; an acquisitive capitalism as an incentive for efficient production and distribution; and its own social and cultural traditions. Moreover, the fact that the British lease on the Hong Kong New Territories expires in 1997 puts additional pressure on the Chinese to modernize. Whether and how it can absorb and integrate such a bla-

---

*In view of the number of times that this country has changed its name — from Cambodia to the Khmer Republic in October 1970, to Democratic Kampuchea in January 1976, to the People's Republic of Kampuchea in January 1979 — I have followed the practice of sticking to one form, Kampuchea, throughout.

tantly individualistic and highly internationalized entity as Hong Kong will have far-reaching implications for the future of the region. As the centre of business activity for the overseas Chinese, with its tentacles reaching into both the communist and the capitalist worlds, Hong Kong is far more important than mere economic statistics suggest. Moreover, the way China handles the Hong Kong question is bound to have an effect on the fate of Taiwan and its population of 18 million ethnic Chinese.

*   *   *

Ever since it opened its doors to the modern world in 1868, Japan has suffered from a sense of being alone in the community of nations, as well as of being isolated from the mainstream of events. Geographically far away from what were then the centres of power, and with a modernization process very different from that of other countries, it was difficult for Japan to be included among the group of Western advanced countries, while there was practically no community of interest with other Asian countries to compensate for this lack. In every sphere, the dichotomy between the West and Asia continued to be a problem, and Japan remained a loner, not really part of any international system.

At the beginning of the twentieth century, Japan tried to plug into the British-led international system through the Anglo-Japanese alliance (1902) and, using this strength, was able to overcome the crisis of the Russo-Japanese war (1904–5). However, as the power of the United States replaced that of Britain in the Pacific after World War I, the Anglo-Japanese alliance was dissolved and Japan, without any international support, had to combat the often whimsical policies of the United States in Asia — policies which were largely hostile to Japan. As relations with the United States deteriorated, Japan, in desperation, turned to the Axis powers of Germany and Italy, but this did little to bolster its position in Asia. If anything, its isolation increased.

At the same time, Japan tried to create its own regional system by colonizing Korea, Taiwan and Manchuria, and in the 1940s put forward the concept of the 'Greater East Asia Co-prosperity Sphere', embracing the whole region. However, in practical terms this was an empty slogan, a product of wishful thinking; Japan could offer no form of political or cultural leadership with which to attract the countries of the region, only the raw power of military force. Nor has it ever succeeded in establishing a division of labour which would have facilitated the region's economic integration.

After World War II, Japan found itself completely cut off from mainland Asia, and placed under American surveillance and protection as a sort of prisoner-cum-protégé. The policy of the Allied powers at that time was to deny Japan any involvement in continental Asia and rather to let it live in virtual isolation. In this context, the post-war position of Japan was quite different from that of West Germany, which was included from an early date in multilateral endeavours to rebuild Western Europe. In the emerging situation in Asia, there was no longer the kind of vacuum of power or political instability which had induced Japan's advance before the war. The Korean peninsula was now heavily guarded by two Koreas, both armed to the teeth. The new government in Beijing was not about to let Manchuria go without putting up a major fight.

As a result, Japan had to start the process of its post-war reconstruction virtually alone, although it leant heavily on the protective arm of the United States which, apart from its role as a supervisor/custodian, recognized Japan's strategic importance in the Western Pacific. In fact, Japan's post-war relations with the world were developed primarily under the direct or indirect auspices of the United States, as the San Francisco Peace Treaty, various reparation agreements and its mode of entry into a number of international organizations attest.

Japan's spectacular economic success during the 1960s did not solve its problem of isolation. If anything, it magnified its vulnerability and reinforced its dependence on the United States. That was at least part of the reason why it was so seriously unsettled by some of the events which descended upon it during the 1970s, such as the Sino-US rapprochement, the Arab oil embargo and the sudden outburst of anti-Japanese demonstrations in Southeast Asia, which exposed the basic weakness of its position in the world. By good fortune as well as sheer hard work, Japan somehow managed to emerge from these 'shocks' relatively unscathed, with a bigger stature, leaner and more efficient economy, and a set of cordial and businesslike relations with a number of countries in the region. However, the basic problem remains intact, and therefore if Japan is to play its rightful role in the future, regionally and globally, it will have to be able to define its position clearly in the world.

The question for the coming years will be whether and how Japan can reinforce its international position and status while remaining essentially an economic power. In spite of its position of acute insecurity, Japan's commitment to peace is still quite pervasive, and its avowed aim is to remain 'a major power without major military power'. Will such a unique stance continue to be viable for a nation of Japan's size and influence in the increasingly complex regional environment of the 1980s and beyond? Will Japan be able to develop a workable system in the region whereby it can play its rightful role while being able to protect its own political and economic interests? Will it be able to find a goal which is congruent with that of the region and thereby be able to change its essentially 'private' pattern of behaviour? These are some of the questions to which this book addresses itself.

* * *

Chapters 2, 3, 5 and 6 follow in roughly chronological order the main changes in the region's political environment since World War II, and examine the response of individual countries in the region, including Japan, to these events. Chapter 4, 'Japan in the 1970s', focuses specifically on Japan, looking at the events that shook the country during that decade, and attempts to analyse why, for all the shocks and turmoil of these years, it was able to maintain its balance and momentum of growth. Chapter 7 is an overview of the economies of the Asian Pacific region, which has been described as the growth centre of the world of the 1980s, and examines the extent of Japan's involvement. Finally, Chapter 8 considers Japan's position in general and addresses the question of how it is trying to relate — in political, economic and security terms — with the outside world.

Chapter Two

COLD WAR IN ASIA: THE FORMATIVE YEARS

With three out of four of the divided countries of the post-war world in its midst, it was inevitable that the Asian Pacific region became the theatre of a series of 'cold' as well as 'hot' wars which were fought as part of the fierce competition between the communists and non-communists for power and influence around the globe. In Europe, the cold war developed along the boundary — hastily agreed upon shortly before the end of World War II — separating the Soviet-dominated countries and the West. Even before the fighting was over, the Soviets were busy trying to consolidate their position behind this border, which came to be known as the Iron Curtain. By placing in a position of power either the communist party or some faction of it which was particularly receptive to Stalinist rule, they were able to capture country after country, and in the process they deployed every textbook technique of revolution, from propaganda and lies to intimidation, selective terrorism and liquidation.

However, the Russians, undoubtedly aware of the immense might of the United States as against their own vulnerability at that time, never tried to press beyond the Iron Curtain or to extend their power into Western Europe. Their caution even went so far as to check certain tactical advances which the communist parties in Western Europe wanted to make, for fear that such tactics would scare the Allies into direct action against the Soviet Union. When they staged the Berlin blockade in 1948, partly in protest at the planned unification of the Western sectors of Germany and partly in the hope of taking control of West Berlin (which was becoming increasingly embarrassing to them as a showcase of 'freedom and democracy'), they were careful never to disrupt the Allied airlift operations.

Western Europe, similarly, was too scared to meddle with the dividing line, lest it provoke the Soviets into war. Even when opportunities to counteract Russian initiatives presented themselves in a series of upheavals behind the Iron Curtain, such as the semi-detachment of Yugoslavia, uprisings in East Berlin, Hungary, Poland and Czechoslovakia, as well as the latest convulsion in Poland, the West refrained from extending any substantial assistance to the victims of obvious acts of oppression. It can perhaps be argued, therefore, that the peculiarly sterile character of the cold war in Europe stemmed from a curious marriage between the caution and restraint of the Soviets and a deep-seated aversion on the part of the Western nations to the devastation that another war would bring.

In contrast with the situation in Europe, the Russians seem to have had little understanding of, let alone control over, events in Asia. The dogma and strategy of agrarian revolution by means of which Mao Zedong led his party to a stunning victory in China were quite outside, if not alien to, the Marxist-Leninist concept of proletarian revolution.[1] In fact, the Russians even tried to check the advance of the Chinese, who were about to cross the Yangtze river in the spring of 1949, while maintaining their recognition of the Kuomintang government until the last moment. Therefore, the Chinese owed little to the Russians in the events leading to the proclamation of the People's Republic of China in Beijing in October 1949 — an event which one might have thought would have altered the global balance of power decisively in favour of the communists, including the Soviet Union.[2]

The rest of Asia was in a state of complete flux, with a surge of Marxist-cum-nationalist revolutions threatening the last vestiges of Western colonial power or the fledgling systems of the newly independent indigenous governments. The Russians appear to have tried to take some initiative in these revolutionary movements, or at least to coordinate them in such a way as to maximize their damaging effect on the remaining centres of colonial power. The Conference of Youth and Students of Southeast Asia for Freedom and Independence, which was held in Calcutta in February 1948, was one such effort. It seems to have put new life into the guerrilla-type struggles in the Filipino *barrios* and Malay jungles and to a series of armed uprisings from Telengana in India to Madiun in central Java. However, it is also true that these uprisings were set in motion by indigenous forces, with little direct help from the Russians.

In Vietnam, following the Japanese surrender, the Viet Minh Front of the Indochinese Communist Party moved swiftly to take over Hanoi, Hué, Saigon and other major cities, and on 2 September 1945 proclaimed the provisional government of the Democratic

Republic of Vietnam. Although the further advance of the Viet Minh forces was checked by the return of the French (which the Soviet Union did not try to prevent because its priority in Europe at this point was to maintain friendly relations with France), the revolutionary dynamism thus accumulated was to wreak havoc in the region later.[3]

The Korean War

The one exception to the Soviet Union's policy of restraint was in the Korean peninsula, where there was a clear demarcation line along the 38th parallel between the Russian and the American spheres of influence. It was a situation very similar to that of the Iron Curtain in Europe, and the Russians followed more or less the same tactics of solidifying their position behind the 'line' by establishing a communist government under Kim Il-sung. It is therefore difficult to understand why, on 25 June 1950, they suddenly decided to give the go-ahead to the North Koreans to cross the parallel and move into the South. It was the first full-scale military operation which the communists had embarked upon since the end of World War II and, as such, was an extremely risky move, running the danger of provoking the all-out retaliation of the United States.

Admittedly, the United States had withdrawn its forces from South Korea in June 1949, and had indicated on a number of subsequent occasions its reluctance to keep its armed forces in the peninsula; this may have precipitated the Russian decision.[4] Also, by this time, China had secured complete victory over the entire mainland, reinforcing the sentiment, particularly in Asia, that communism was the wave of the future and its victory inevitable. The Soviet Union may have thought that it could ride on this tide and expand its position in the entire Korean peninsula. It is also possible that, being Western nations, both the United States and the Soviet Union were inclined to take rash and impetuous actions in Asia that they would not have taken in Europe. But lack of understanding of Asia's political and cultural complexities may not be the only explanation. It could also be that the ravages of the war in Asia would provoke only an indirect, and therefore less painful, psychological impact upon themselves, as compared with actions in Europe.

In any case, the Soviets' assumption that the United States would not intervene proved to be totally wrong, and by September 1950 they had been penalized for their misjudgement by losing their position in almost the whole of Korea, with the US forces triumphantly marching towards the Manchurian border. When General MacArthur met President Truman on Wake Island on 14 October 1950, he proudly assured him that the victory had been won in both North and South Korea, that all resistance would end by Thanksgiving Day and that the US troops would be able to return home by Christmas.[5]

The situation changed abruptly, however, in less than a month, when the Chinese intervened — at first stealthily and then openly — with a massive army of 200,000 'volunteers'. The Chinese action altered the entire political and strategic dimension of the war, elevating it automatically from a regional conflict to a struggle among major powers, with the distinct possibility of further escalation to global war. Neither the United States nor the Soviet Union seemed to have any idea at that point how to contain the situation, which was rapidly getting out of control. On 30 November President Truman announced, perhaps in desperation, that there was 'active consideration of the use of the atomic bomb' to meet the military situation in Korea.[6]

Frightened by the sudden prospect of a major war, the British prime minister, Clement Attlee, flew to Washington to appeal to the United States to contain the conflict and to make sure that it did not spill over into Europe. As the price for a cease-fire, the British counselled the United States to offer the Chinese a seat in the United Nations in place of the Kuomintang government in Taiwan. Attlee, whose government had recognized the communist government in Beijing earlier, argued that China and the Soviet Union were natural rivals in the Far East, and that therefore the Western aim ought to be to exploit the differences between the two. According to British thinking, China was a potential candidate for Titoism. Truman was of no such opinion, saying that the Chinese communists were Russian satellites, and that he saw the problem as part of a pattern: namely, 'After Korea, it would be Indochina, then Hong Kong and then Malaya' — a statement that unwittingly projected a prototype of the 'domino theory' which was to become one of the basic tenets of American policy in Asia.[7]

The conversation also revealed the divergence in perception between Britain as a regional power and the United States as a superpower. The British argued that Europe was more central than Asia to the East-West confrontation, and therefore that the war in

Korea should not be allowed to escalate into mainland China because it would expose Europe too much to the danger of Soviet intervention. The United States naturally agreed that the war should not be allowed to escalate, but pointed out that, in a broader context, the United States could not adopt an isolationist position in the Pacific, ignoring what the Chinese communists had been doing, while at the same time taking a strong anti-isolationist stand against the threat of communism in Europe. This was perhaps a gentle reminder that the United States, which was responsible for the defence of the whole world, could not permit its policies to be determined by the preferences of smaller regional allies.[8]

By January 1951, the United States and its allies had regained sufficient strength to roll back the Chinese to a battle line not far from the 38th parallel. At this point the war became static — an impasse which lasted until July 1953, when an armistice was reached after some two years of negotiations. It is significant that, throughout the entire war, the Russians had themselves managed to stay away from the battlefield, leaving the Chinese to bear the brunt of the fighting, almost single-handed and at very great cost, against the world's most sophisticated military machine.

The Chinese certainly had a compelling reason to fight the war in Korea. When the North Koreans collapsed under the US counter-offensive, the Chinese had to face the uncomfortable prospect of a massive American military build-up along the border with North Korea. Given the vociferous anti-communist rhetoric that was rapidly gaining strength in the United States, there was even the possibility that the United States and Taiwan would mount an attack on the mainland, capitalizing on the Chinese predicament in Korea. None the less, the Chinese were made to pay an exorbitant price for their initial misjudgement of the US response, an error for which the Russians were primarily responsible. The story that China had to pay with commercial loans for every weapon it received from the Soviet Union during the war may suggest that, by making the Chinese bleed in Korea, the Russians hoped to make China more dependent on the Soviet Union — one way of dealing with China's potential influence as a rival centre in the communist world.[9]

The United States was not unaware of these difficulties in the Sino-Soviet relationship. The China White Paper, published by the State Department in August 1949, in fact presented conflicting views of China and its future course: on the one hand it saw China as a potential Yugoslavia; on the other, it tended to write it off as a satellite of the Soviet Union. The speech made on 12 January 1950 by Dean Acheson, then Secretary of State, in which he effectively excluded Taiwan and Korea from the US defence commitment in the Far East (and which has been interpreted as one of the factors that precipitated the North Korean attack across the 38th parallel) may well have been an expression of a wish on the part of the United States to see China move further in a Titoist direction. However, with the signing of the February 1950 Sino-Soviet Mutual Assistance Treaty (a *de facto* military alliance in which the United States and Japan were regarded as potential enemies), the 'Yugoslavia syndrome' interpretation gradually gave way to the view that 'China was lost'.

This view was greatly reinforced and accelerated by the tenacity displayed by the Chinese on the Korean battlefield, which impressed upon the United States communism's sinister ability to transcend national and ethnic boundaries. There was now little doubt in American minds that communism was on the march, determined to conquer the world. It strengthened the conviction of Americans that theirs was the only country which had the power, responsibility and right to defend the free world and maintain international order.[10]

The long record of friendship between the United States and China before World War II aggravated the USA's indignation at seeing the Chinese 'being forced into the battle' against it.[11] Since shortly after the turn of the century, the United States had regarded China as its friend-cum-protégé, whose fledgling 'democracy' needed the support and attention which only the United States could give it. Seized by a typically American moral fervour, a whole army of missionaries and volunteers went to China to work in schools, churches and hospitals throughout the country. Meanwhile the State Department tried to make the defence of China's 'territorial integrity' against the aggressive imperialists, notably Japan, a central thrust of its Asian policy.[12] Resentment at seeing once friendly and mutually rewarding relations being twisted and betrayed by a hostile ideology was one of the factors behind the USA's subsequent commitment to the 'containment of China' — a policy which was to have a tremendous influence on the political landscape of the Asian Pacific region for many years to come.

The Polarization of Japan

As the cold war set in in the Far East, it became necessary for all the countries of the region to show clearly where they stood. For countries like South Korea and Taiwan, a commitment to the US-led forces of 'freedom' was the only real choice. Moreover, for South Korea, which had just survived the ordeal of a fratricidal war, this was what the people themselves wanted. For these two countries, anti-communism as a basic principle has long dominated political actions both internally and externally.

For Japan, which had been occupied effectively only by the United States (there was a small British Commonwealth force, but no Soviet force), there was no other choice but to make its commitment to the United States, and, in signing the San Francisco Peace Treaty in September 1951, it aligned itself clearly with the free world. Economically, too, it made obvious sense. Moreover, this coincided with the feelings of the majority of Japanese, whose fear and dislike of communism were in part the legacy of a strong anti-communist orientation before the war. Even more, however, they were angered by the way the Soviet Union took advantage of Japan's defeat at the end of the war to seize control of the islands north of Hokkaido and to send more than a million captured Japanese to suffer up to eleven years of forced labour in the Siberian gulags.

Nevertheless, the cold war was to have an unexpectedly complex influence on Japan's internal politics. In fact, as though it were the price it had to pay for the good fortune of not being divided, Japan discovered its own domestic polity sharply divided along the ideological fault-line which seemed to be sending the whole world to the brink of World War III. This polarization stemmed partly from the post-war psychology of the Japanese, partly from the changing attitude of the United States, which was then playing the role of all powerful guardian-cum-warder of Japan, and partly from the growing polarization of the world itself.

After the war, the Japanese were understandably filled with remorse, not only on account of their part in the war itself, but because of a pre-war system which had allowed the military to dominate the country, only to bring about its eventual destruction. There was a near-unanimous sentiment that this should never be permitted to happen again. The policy and attitude of the US authorities, at least at the initial stage of their occupation, as symbolized by the 1947 'peace constitution', were largely in harmony with this line of thought in Japan. Admittedly, there were fairly harsh guidelines, issued by the Far Eastern Commission,* aimed at teaching Japan a permanent lesson, such as holding down its future living standards to 1930–4 levels. Nevertheless, many of the changes which the United States had forced upon Japan were regarded by the majority of Japanese as liberation rather than oppression.

Many of the communists who had been imprisoned were released, left-wing political parties, including the Japan Communist Party (JCP), were granted freedom of action, and the rights of trade unions were guaranteed. The military establishment was dissolved completely, and the police, together with the Home Ministry, were decentralized; the base of political participation in local government was broadened. Further drastic measures, such as the seizure and redistribution of large agricultural properties and the dissolution of the *zaibatsu* (the great family-held industrial and financial combines), were implemented in line with the basic tenet of the 'democratization' of Japan.

These policies accorded with the image of the United States as champion of the global crusade against fascist dictatorship. Mao Zedong stated, in a political report prepared for the Chinese Communist Party in April 1945, that the relationship of the United States, the Soviet Union and Britain was one of unity among anti-fascist democratic forces, and a Chinese communist publication in 1944 described the United States and the Soviet Union as the 'twin stars of the democratic world'.[13]

However, by the time the Truman Doctrine was proclaimed in 1947, reflecting an intensification of the confrontation between the Russians and the Americans, the policies of the Occupation authorities were beginning to shift — moving towards treating Japan as a political ally in US global strategy. In addition, the need to alleviate the burden on the American tax-payer argued in favour of accelerating the pace of Japan's economic recovery. Other factors, too, caused the Occupation authorities to re-evaluate some of their 'democratization' measures. Thus, in 1947 they banned a general strike planned by a wide variety of unions across the country, and in the following year they regulated the rights of public enterprise workers to strike and

---

*An inter-allied body set up to formulate policy for the occupation of Japan. Established in December 1945 by the United States, Britain and the Soviet Union (with the concurrence of China), it replaced the Far Eastern Advisory Commission. It met regularly in Washington from early 1946 until September 1951, when the San Francisco Peace Treaty was signed.

made extensive changes in the programme for dissolving the *zaibatsu*.

These moves were welcomed by the conservative elements in the country, but in labour circles and the left-wing parties they bred a sense of disillusionment and betrayal at the hands of the Americans. Members of these leftist groups managed to mobilize considerable public support under the banner of the peace movement, whose goal of 'democratizing' Japan was perceived as being thwarted.[14] With remarkable skill, which perhaps they had learnt from classic popular-front tactics in Europe, they managed to draw on a wide range of society, which included journalists, trade unionists, workers, intellectuals, students and housewives. In this, they were able to rely heavily on the pacifism which by then had taken very deep root in Japan.

Also, there was a widespread fear among ordinary citizens that, in its zeal to fight communism, the United States would facilitate the return to power of the reactionary elements that had been guilty of causing the war, and that once again the country would be steered towards nationalism-cum-militarism. This fear was so prevalent that even such terms as 'patriotism' or 'loyalty to the nation' came to be suspect. Moreover, the so-called 'two camp' theory, newly propounded by the Cominform, helped to give such arguments global authority. According to this theory, there were two forces in the world, imperialist and democratic, or reactionary and progressive, with the former being represented by the United States and the latter embracing the revolutionary forces around the world under the 'progressive' leadership of the Soviet Union.

When the war broke out in Korea, and the bulk of US forces stationed in Japan had to be moved out to the battlefield, the United States began seriously to consider the rearmament of Japan. General MacArthur requested that a 75,000-strong National Police Reserve* with a paramilitary capability be set up to take the place of the departing US forces in maintaining public order. And, subsequently, in 1951, when negotiating with Japan on the prospective terms of the Peace Treaty, the US envoy, John Foster Dulles, urged strongly upon Prime Minister Yoshida Shigeru that Japan should build an army of up to 350,000 men.[15]

These requests met with considerable dismay in Japan. For one thing, they seemed to be in blatant contradiction of Article 9 of the

---

*In 1952 its name was changed to the National Safety Forces, and in 1954 these were reorganized as the Self-Defence Forces (SDF).

Japanese Constitution (authorized, if not authored, by the Occupation administration), which stipulated that Japan renounce the right to go to war or to maintain armed forces for that purpose. For another, it was regarded as the first step towards a breach of the nation's near-sacred pledge never to rearm. Also, it was feared that, once given a bridgehead, however small, the newly revived military would find ways and means to expand, at gunpoint if necessary, until it again led Japan along the road to disaster.

Another aspect of the USA's policy about which Japan had grave misgivings was its expressed aim of containing China. The American argument was that communism was a dangerous epidemic with the potential to afflict country after country, threatening the very foundations of democracy. Therefore the forces of communism had to be either destroyed or contained. The Japanese, for their part, although they saw communism in the same light, felt that its symptoms varied according to the political culture and tradition of each patient, so that one must deal with it sensitively and selectively, with the aim of averting the danger and steering it in a less dangerous direction. Japan was not alone in taking this stand; other American allies, notably Britain, thought along similar lines. Nor was it confined to the opposition in Japan; it was a view held widely in government and business circles, and by Prime Minister Yoshida himself.

Therefore, when the United States tried to persuade Japan to accept the regime in Taiwan as the legitimate government of China, Japan showed some resistance. Again it was Dulles, at a meeting with Yoshida on 18 December 1951, who forced Japan to accept the American position. As a special envoy of President Truman, he was adamant that the Japanese should make their position clear, hinting that without a definitive expression of Japan's intent, the San Francisco Peace Treaty, which had been signed in September that year, might well fail to be ratified by the US Senate. Understandably, Japan capitulated (despite the express support of Britain), and the issue of recognition was settled through the 'Yoshida letter' of 24 December 1951.[16] This course of events left a suspicion in the minds of the Japanese that American policies towards Asia were not always based on a consensus among the allies, and that they were not always either in Japan's national interest or in the interest of regional stability.

These issues turned out to be timely and welcome ammunition for the opposition, whose influence was being constantly eroded by internal divisions. Ever since its inception, the Japan Socialist Party (JSP), the largest opposition party, had been split by differences in ideology and tactical orientation. The trade unions, too, embraced a wide spectrum of opinion and were perennially afflicted by fierce competition between the communists and their fellow-travellers on one hand, and those who refused to give in to their leadership on the other. The questions of rearmament and China were always helpful to them in overcoming these differences and rallying their forces alongside the mass of the people, students, intellectuals, journalists and ordinary citizens, in anti-government and anti-US activities.

Thus, the 1950s turned out in Japan to be a decade of unprecedented political polarization. Many key industries, in particular coal mining, shipbuilding and electronics, were repeatedly paralysed by highly politicized strikes and industrial action, and campaigns against the US military bases in the country became a feature of the politics of the day.[17] With a peculiarly self-imposed sense of mission to defend Japan from remilitarization, the majority of the newspapers and other media freely branded the conservative governments as reactionary, and attacked their subservience to the United States as being detrimental to the national interest.

When Prime Minister Kishi Nobusuke acceded to power in February 1957, the unrest in the country came to a head. Every new piece of legislation which the Kishi government attempted to get enacted, whether an amendment to the Police Duties Law or a proposal for a new system to rate the efficiency of teachers, was considered to be part of a sinister plot to turn the clock back to the pre-war era, and was countered by mass demonstrations.

The climax of these struggles was reached when the ratification of the revision of the US-Japan Mutual Defence Agreement (often called the MSA or Mutual Security Assistance Agreement) was presented to the Diet in 1960. This had been agreed and signed by the US government earlier in the year at the request of the Kishi government. The aim of the amendment was to rationalize the security relationship between the United States and Japan, and to protect Japan's interest more specifically than the original Security Treaty of 1951, which had been hastily agreed upon at the time of the San Francisco Peace Treaty. However, in the highly volatile environment of spring 1960, the content of the revision was not really at issue: the opposition was determined to resist the treaty in any shape or form.[18]

Under the banner of the People's Council for Preventing Revision of the Security Treaty, wave after wave of demonstrators charged the Diet buildings, reinforcing the battle which was taking place inside, where the government party, in clear majority, would automatically ratify the treaty. The opposition members often used sheer physical force to prevent the chairman of the committee in charge, or even the speaker of the Diet, from taking their seats to announce the voting. On 19 May, a month before the planned visit of President Eisenhower to Japan, an exasperated government took the drastic measure of bringing the police force into parliament and having all the opposition members removed from the chamber. The vote was thus pushed through by the government party, the Liberal Democratic Party (LDP), alone.

Legally the government's action may have been valid, since the LDP had a comfortable majority in the lower house. However, to the demonstrators and their sympathizers it was a red rag to a bull. Having been defeated on the ratification of the treaty, they were now determined to stop the presidential visit, if only to humiliate the government and let Kishi know that he could not get away unscathed with victory on both counts. On 10 June, members of the well-known Zengakuren student movement mobbed the car of Eisenhower's press secretary, James Hagerty, on his arrival at the airport for preparation of the presidential visit. On 15 June, when a massive crowd broke down the gates by force and entered the grounds of the Diet, one girl student was killed. The next day, Kishi made the painful decision to cancel the Eisenhower visit and himself to resign.

There is little doubt that the struggle surrounding the Security Treaty epitomized the acute political division that prevailed in Japan in the 1950s, and that it reflected a mounting ideological tension around the world in general. However, it is also true that the moving force behind this campaign was essentially internal, without much incitement or support from outside, communist or otherwise. The fact that such a tremendous wave of mass reaction died down almost the instant that Kishi resigned suggests that, in part, it was a trial match in domestic politics — a test of will and strength as to how far the government's reactionary activities could be checked and prevented by massive extra-parliamentary action. One can perhaps argue that it was an experiment in the dynamics of internal politics for a people who had acquired through defeat and the Occupation a completely different type of political system: the demonstrations were, in effect, a confirmation of that system's operational viability.

It is ironic that the Japanese could afford to indulge in such an extraordinary and basically self-centred experiment at this juncture. They could do so precisely because the strength of the US commitment, as embodied in the Security Treaty (the very issue that lay at the heart of the country's unrest), gave them the necessary guarantee against external threat. It was the kind of luxury which would not have been allowed to a country like West Germany, located right at the centre of the East-West confrontation, for fear that it would upset the precarious balance between the two superpowers. (The analogy with West Germany is discussed in detail in Chapter 8.) One can perhaps argue that, having been completely cut off from continental Asia, post-war Japan was somewhat detached from the area of direct contention between the superpowers. The Russians seem to have had little intention of tampering with a country which was obviously in the US sphere of influence, and therefore for the time being the United States was prepared to sit back and watch, if apprehensively, the wave of anti-US demonstrations engulfing Japan without feeling the need to intervene.

Japan reaped some unexpected benefits from experimenting with the working pattern of its own 'democracy'. Since massive demonstrations seemed to be shaking the country's political foundations, both China and the Soviet Union appeared to see an advantage in being conciliatory to Japan in the hope of encouraging anti-US sentiment. The Soviet Union agreed to normalize its relations with Japan in 1956, and China repeatedly expressed its support for the 'people' of Japan in their anti-imperialist struggle. As a result, Japan was able to develop trade relations with a number of communist countries, notably China, with which it had no inter-governmental relations. As for the United States, it took no measures to deprive Japan of such a blatantly 'economistic'* activity as trade with an unrecognized communist state, for fear that this would aggravate the anti-US feeling in the country.

The Emergence of Economism

Ikeda Hayato, who in July 1960 succeeded Kishi as prime minister, chose to lower the political temperature of the nation by proposing an essentially 'economistic' policy framework. He placed economic development at the centre of his programme and asserted that Japan should aim at establishing its international status by means of its economic strength.[19] Also, he proclaimed 'low posture' as the tactics of

---

*The term 'economism' and the coinage 'economistic' are used to denote an attitude that gives priority to economic, rather than political, considerations.

day-to-day politics, which meant avoiding controversy, ideological or otherwise, as far as possible. The new government's cautious, 'safety first' attitude was to invite criticism from various quarters inside and outside the country, but the people themselves seem in general to have welcomed it. Understandably, after many years of persistent ideological conflict, they were suffering from 'confrontation fatigue'. Even though they had been willing to experiment with the newly acquired 'democracy' at the time, the confrontational politics of the 1950s were probably alien to Japan's strongly consensus-oriented political culture.

As part of his economic programme, Prime Minister Ikeda put forward a scheme for 'doubling national income in ten years', which succeeded in rallying the support of the population and further checking the momentum of anti-establishment forces. The trade unions gradually shifted their platform from political to 'economistic' demands, while the radical members of the Zengakuren movement passed into conventional society.

It was fortunate for the Ikeda government that the United States did not raise any strong objections to its economism. The United States would have liked to see Japan commit itself once and for all to its crusade against communism. The Kishi government appeared to have done this, but the outcome of Kishi's policies had been as alarming to the United States as to Japan. The cancellation of Eisenhower's visit to Tokyo in June 1960 had been a blow to its international prestige; it coincided with the anti-US demonstrations that marked Vice-President Nixon's visit to Latin America. The television pictures from Japan, showing clashes between the police and students in front of the Diet, made it look as if the country were on the eve of revolution. As a result, the United States accepted Ikeda's 'low posture' stance as second-best choice, and supported his endeavour to stabilize the country.

This gave the Ikeda government an opportunity to experiment with 'low posture' politics without fear of provoking US repudiation. On the question of defence, for instance, Ikeda refused to recognize Japan's need to depend on the US guarantee. He asserted, during the November 1960 election campaign, that the ultimate security of Japan lay with the United Nations, and that the US-Japan Security Treaty was merely a temporary measure until such time as the United Nations began to function satisfactorily. This somewhat evasive logic paid off, and the LDP was returned to power with an increased majority.

Ikeda showed similar caution over the return of Okinawa, which could easily have developed into an explosive issue between the two countries. He knew that, as long as the United States was committed to the 'containment of China', there was little likelihood that it would return this strategically important group of islands to Japan. Therefore, instead of demanding its return, Ikeda chose to offer an increase in Japan's contribution to Okinawa's social and economic development. By this 'economistic' method, he was able to avoid another surge of anti-US feeling in the country, and at the same time to expand Japan's participation in the administration of Okinawa and prepare the ground for future negotiations for its return.[20]

When Ikeda visited Washington in June 1961, the Kennedy administration agreed to set up a Japan-US Joint Committee on Trade and Economic Affairs. A joint communiqué was issued, and it was proposed that the committee should meet annually, alternately in Japan and the United States. Already anticipating a showdown with China in Indochina, the United States seemed anxious to keep Japan on its side. The membership of the committee was to include six persons from each side, including Secretaries of State, Interior, Treasury, Agriculture, Commerce and Labour from the United States, and Ministers of Foreign Affairs, Finance, Agriculture, Forestry and Fishery, International Trade and Industry, Labour, and Economic Planning from Japan. By committing itself to a forum of dialogue at such a high level and on so large a scale (a scale of operation seen only once before, in the case of Canada), the United States demonstrated the importance it attached to Japan. It also conveyed its support for Japan's economistic stance by agreeing to confine the committee's discussions to 'bilateral economic issues'.

Surprisingly, the committee met a total of nine times during the twelve years of its existence: from the initial meeting in Hakone, Japan, in November 1961 through to July 1973. This coincided with the most traumatic period in American post-war history, with the three administrations (Kennedy, Johnson and Nixon) largely occupied by the war in Vietnam. Nevertheless, the United States continued faithfully to attend these meetings, with its full membership from the Secretary of State down — a reflection of Japan's importance in its regional policy calculations. Even the planned November 1963 session was postponed by only two months as a result of the Kennedy assassination.[21]

In the earlier meetings, Japan was primarily concerned with redressing the chronic deficits in bilateral trade with the United

States, which ran to the tune of an annual $800 million, and urged it to modify some of the clauses of its 'Buy American Act'. It is interesting to note that, at that time, the United States countered Japan by preaching that the trade imbalance should be looked at not solely on a bilateral basis but as a multilateral problem, involving the balance with other countries as well — the exact reverse of the two countries' positions today. For all the talk of a multilateral solution, however, there was little Japan could do about the CoCom* restrictions on trade with communist countries, the discrimination against Japanese goods in Europe and the shortage of foreign exchange in other markets, notably Southeast Asian countries. Therefore, the United States agreed to give active support to Japan's application for membership of the Organization for Economic Cooperation and Development (OECD) as well as its request for removal of discrimination in Europe.[22] As a result, in 1963 Britain and France rescinded the use of Article 35 of the General Agreement on Tariffs and Trade (GATT) against Japanese imports, and in 1964 the OECD accepted Japan as a member.

In retrospect, the committee seems to have had the effect of shielding Japan from the political realities of Asia during the 1960s. Even when the Asian Pacific region was embroiled in the everescalating war in Vietnam, Japan continued to avoid political issues in its discussions with the United States in the committee. It was as if Japan wanted to apply to its relationship with the United States the principle of *seikei bunri* (the separation of economics from politics), which was originally conceived in the 1950s as a device to get round political and ideological difficulties over Japan's trade with China. In fact, under the guise of an economistic expediency, the concept of *seikei bunri* represented a widespread scepticism in Japan about the ultimate efficacy of cold-war premises. Although the United States was not at all in the mood to appreciate views of this kind, it refrained from pushing political discussions in the committee, primarily in accordance with the spirit of the Ikeda-Kennedy communiqué. Perhaps this distanced the US-Japan relationship from realities, and it may have helped the two nations to remain on good terms while pursuing different goals.

---

*The Coordinating Committee for East-West trade, established in November 1949, which coordinates the security export control policies of the NATO countries, less Iceland, plus Japan.

The Vietnam War

War is an exercise which seldom produces the results that the contestants had bargained on. Vietnam was no exception. At the end of the war, which had been designed primarily to contain China, the United States had to engineer a spectacular comeback for China as a dazzling new star in the international firmament.[23] The war which was to have thwarted the advance of communism in Southeast Asia, in fact led to the entry of the Soviet Union into the region in its new role as a mentor/patron of Vietnam. Moreover, a war which was intended to fortify the morale of the free world ended up undermining the relevance of the cold war itself. Nor were the disappointments confined to the United States. The Vietnamese, who were to have joined 'the ranks of the vanguard nations of the world',[24] soon found themselves caught up in the internecine struggle of the communist giants.

Nevertheless, in spite of disillusionment on both sides, the war in Vietnam had a tremendous influence on the political environment of the Asian Pacific region. The way the United States sent half a million troops into the tiny southern half of Vietnam, and then pulled out essentially of its own free will, sent a series of shock waves through the region, activating a fundamental realignment of relations among the nations, including China.

In retrospect, there were two areas in which the United States had difficulty in carrying out its designs in Vietnam. First, it seems to have failed to make an up-to-date assessment of the nature and strength of its ultimate opponents, namely the Soviet Union and China, which supposedly formed the nucleus of the so-called 'monolith' of communism. Second, it had difficulty in pursuing its global objectives while trying to address itself to all manner of problems which were peculiar to the South Vietnamese situation.

The Sino-Soviet relationship was deteriorating already in the summer of 1960, when the Soviets abruptly discontinued their technological assistance programme for China and withdrew all their engineers and advisers. In July 1963, a top-level meeting on ideological issues which took place between Deng Xiaoping and Mikhail Suslov, representing the communist parties of the two countries, ended in a complete rupture, showing what an illusion the 'monolith' of communism had been.[25] In October 1964, China exploded its first nuclear device — a show of force which was aimed as much at Moscow as at Washington, although the latter took it extremely seriously as further evidence of Chinese belligerence.

During the mid-1960s, when the United States was becoming increasingly involved in the war in Vietnam, China, in the name of the Cultural Revolution, was heading steadily for the mortal internal rift which placed the all-important party organization in serious disarray, even risking a complete collapse of order. The conflict was partly ideological and partly personal, between Party Chairman Mao Zedong on the one hand, and a group of his high-ranking comrades, such as Liu Shaoqi and Deng Xiaoping, on the other.

Throughout his life Mao believed in the legitimacy of the agrarian-based mass revolution, supported by the indigenous radicalism of China's peasant population, which had proved its efficacy in giving the Chinese Communist Party (CCP) a theme as well as the means of transforming China's basically agrarian society into one of the most successful communist revolutions in the Third World.[26] The difficulty was that while Mao held to an almost romantic belief in a continuous process of mass revolution, more pragmatic comrades and party bureaucrats felt that what China needed now, fifteen years after the revolution, was a programme of social and economic development which called for law and order in the land as well as comprehensive and consistent economic policies. The memory of the failure of the Mao-inspired Great Leap Forward of 1958 was still fresh in their minds. Sensing that such differences of opinion might erode his own power base, Mao decided to strike back with his full force, and, by unleashing the raw power of youth in the shape of the Red Guards, he was able to make a frontal attack on the party bureaucracy throughout the country. When and where the party machine showed resistance, he did not hesitate to use the army to subdue it.[27]

In these circumstances, China was in no position itself to fight the war in Vietnam, or even to coordinate its policies with the Soviet Union. In fact, the Soviets suspected from the beginning that the true intention of the Chinese as regards the war in Vietnam was to bleed both the United States and Vietnam so that China would later be able to assert its dominance over the area, while at the same time reducing the threat of a direct attack from the United States. They interpreted Mao's comments to Edgar Snow that 'it is a good thing to have the Americans in South Vietnam', and that 'China would fight only against a direct attack by the United States on her territory', as being tantamount to giving the Americans a free hand in the whole of Vietnam, South as well as North.[28]

Moreover, the Soviets were completely baffled by the volatile nature of the Cultural Revolution and seemed to be at a loss what response to make. They could not understand the sudden surge of revolutionary zeal in China — a bewilderment that probably reflected the extent of bureaucratization in their own society. In August 1966, China intensified its opposition to the Soviet Union by adopting, at the Eighth Plenary Session of the Central Committee, an official decree which stated that the greatest danger for China was not American imperialism but Soviet revisionism.[29] The Soviets in turn began to suspect that what the Chinese really wanted now was to have both superpowers exhaust themselves by fighting over Vietnam, thus echoing their interpretation of Chinese intentions in 1958 when China started shelling the islands of Quemoy and Matsu off the Taiwan Strait.[30]

It can perhaps be argued that if China and the Soviet Union had indeed been the core of a monolithic world communism, with Vietnam as their spearhead, as the United States had assumed, the war in Vietnam would have taken a different course. Given the superiority of its military technology, the United States might have achieved some credible victories. As it was, when the US troops went into Vietnam, the enemies they expected were not there, and they had to fight such elusive and ill-defined targets as the nationalistic fervour of the North Vietnamese and the general dissatisfaction of the people with the government of South Vietnam — targets which they were not equipped or meant to deal with. This leads to the second difficulty faced by the United States: namely, why and how the huge commitment in Vietnam came about.

The difficulty which the United States encountered in Vietnam seems to have stemmed from the region's lack of structure. In Europe, the US commitment was based upon its membership of the North Atlantic Treaty Organization (NATO) which regarded an attack on any one member as an attack on all and called for an automatic response from the United States in the event of Soviet aggression. The corollary to such a comprehensive commitment was the full-scale deployment of an awesome nuclear arsenal, and the semi-permanent placement of 300,000 US troops in Europe, primarily in West Germany. The Europeans were also obliged to increase their own defence capability, and, to help them to generate the economic base for this, the United States poured in every kind of aid and assistance during the 1940s and 1950s, through various arrangements, notably the Marshall Plan.

In Asia, the only countries which could look to the United States for an automatic response of a kind roughly comparable with that of the NATO provision were the Northeast Asian countries of South Korea, Japan and Taiwan (effective until the rapprochement with the mainland in the early 1970s). In contrast with the NATO case, however, these were merely bilateral treaty links. Nevertheless, they seem to have been effective in containing potential conflicts along the Taiwan Strait and the demilitarized zone in Korea. In Southeast Asia, the United States preferred to avoid an open-ended commitment of this kind, even when it was promoting — in 1954 — the establishment of the South-East Asia Treaty Organization (SEATO), which in some respects sought to pattern NATO. In spite of a specific plea from Thailand for a commitment to automatic response in the event of an act of aggression, the United States decided in favour of a low-cost arrangement, which was thought to be enough to deter aggression in this region, keeping its options entirely open on whether to intervene or not.[31] Thus, the United States managed to avoid a so-called automatic commitment, which would have involved an unacceptable risk given the liquid state of politics prevailing in Southeast Asia at the time. Consequently, in the absence of clear strategic guidelines, the United States had to expose itself to a different kind of danger, namely *ad hoc* involvement in political and military situations in any number of countries without any effective mechanism for self-restraint.

Some argue that the escalation of the American commitment in Vietnam was brought about not as a major policy decision but as an incremental combination of specific commitments, such as the bilateral military assistance agreement with South Vietnam, which carried the implication of further support.[32] John MacNaughton, former Assistant Secretary of Defense, takes a more cynical view: according to him, the real reason for going on in Vietnam was 10 per cent to save Vietnamese democracy, 20 per cent to preserve a proper balance against China and 70 per cent prestige — to avoid the humiliation of defeat.[33]

Whatever America's motives, it is probably true to say that the astounding size of its military commitment became necessary as a result of its over-involvement in the internal situation in South Vietnam. In a zealous, yet ill-conceived, pursuit of its global objective, namely the containment of communist movements in Asia, the US field administration in Saigon was inevitably drawn into the complexities of Vietnamese politics. A case in point is the alleged involvement of the United States in the *coup d'état* of November 1963, which brought about the death of President Ngo Dinh Diem.

The policies of the Diem administration, particularly those of the President's brother and confidant, Ngo Dinh Nuh, appear to have been both corrupt and repressive, and the United States therefore urged Diem to remove his brother from a position of power. When Diem would not concur, the United States reportedly began to assist, or at least ceased to restrain, the covert moves of a group of generals who subsequently staged the coup and removed the President. It was a typical example of the kind of dilemma which the United States had to face in the South Vietnamese environment. An attempt to remove an elected president from his position would obviously go against its objective of reinforcing 'democratic' rule in Southeast Asia. However, in view of the perceived urgency in its other purpose, that of containing communism, it was perhaps considered an inescapable part of its programme.

Unfortunately, the generals who staged the coup were unable to govern the country. This left the United States with full responsibility for keeping war-torn South Vietnam moving, as well as thwarting the ever-increasing threat from the communist forces. Before long, President Johnson found himself ordering air attacks on North Vietnam as well as a massive reinforcement of US combat troops in the South. In the words of Frederick Nolting, Jr, former US ambassador to Saigon, this 'turned a Vietnamese struggle into an American war. Morally, it hung an albatross around our neck.'[34]

Perhaps it was this albatross which caused the curious air of illegitimacy that lingered over successive governments in South Vietnam and made the nations of the region wary of extending their full support to the American effort in Vietnam. With the exception of South Korea, which sent in 50,000 combat troops (who fought valiantly alongside the US forces from 1965 onwards), few nations were ready to participate.

Lessons of the War

The massive American involvement in Vietnam, and the subsequent course of events which led to the downfall of the Saigon government in 1975, had a profound impact on the other non-communist nations of the region. As an extension of the American-led 'containment of China' policy, the war forced these countries to put to the test their own commitment to the cold war. They were obliged to engage in some hard thinking about the impact that the war would have on them, and how best, within a limited range of options, to maximize the benefits and minimize the costs.

The most enthusiastic response came from South Korea, which was itself, despite the UN forces securing the demilitarized zone after the Korean war, still under threat from the North. The perception that it was a frontline state in the cold war was held by government and people alike. The policy adopted by South Korea, by which armed forces were sent to Vietnam at the request of the United States, had a broad measure of support within the country. President Pak justified this policy on three grounds: that South Korea had a moral obligation to the free world's collective defence against communism; that it should repay the debt of having been saved by the United States and its allies; and that the commitment of troops was an act of indirect self-defence in that if Vietnam fell, then the whole of Asia would follow.[35]

South Korea began to send its troops to Vietnam in October 1965 and maintained its presence there until March 1973, with 50,000 soldiers at the peak period, which lasted from 1967 to 1972. It is estimated that South Korea lost 4,000 soldiers in 1,170 operations, mopped up 41,000 enemy troops and swept the enemy from an area of 7,438 square kilometres along the east coast of Vietnam. These figures indicate the importance of the Korean presence for the US effort in Vietnam. In return, US military assistance to South Korea during this period doubled, rising to $1.7 billion during 1965 to 1969, as compared with $800 million in the previous five years.[36] This naturally had the effect of strengthening the American commitment to the security of South Korea itself. Thus, in a speech delivered in Seoul in February 1966, Vice-President Hubert Humphrey stated: 'As long as there is one American soldier on . . . the demarcation line, the whole and entire power of the United States is committed to the defence of Korea.'[37] Of course, this commitment reinforced the image of South Korea as a 'cold-war warrior', and although there was criticism that it was being alienated from the Third World, it was probably a rational move for South Korea at that time.

In contrast, Japan, politically at least, did not derive much benefit from the Vietnam war in that it posed a danger of potentially fierce domestic polarization. The Japanese did not particularly want to see the spread of communism, especially since in the early 1960s economic links with the Southeast Asian countries were beginning to expand, and a peaceful, non-communist regional environment was considered very important. However, they had reservations about whether the introduction of large numbers of troops by the United

States was in fact the best method of containing communism, especially in Southeast Asia. It was clear that if a superpower such as the United States overcommitted itself to the extent of using regular forces in substantial strength in Asia, even in the strictly limited theatre of Vietnam, there was a danger of upsetting the global balance of power, whatever local objective might be achieved.

The way in which the United States escalated its involvement in Vietnam, virtually without consulting either its European or its Asian allies, was beyond their comprehension. The Japanese knew from their experience during World War II, both in China and elsewhere, that it was extremely difficult for a regular army to fight effectively against guerrilla forces in the villages and jungles of Asia; moreover, they thought the designation of the guerrilla forces as the vanguard of world communism to be simplistic. In addition, the repressive measures enforced by successive South Vietnamese governments in an attempt to deal with their many internal problems provoked a strong reaction from the left wing in Japan. It merely added force to the cliché that the United States, in the name of containing communism, had a penchant for supporting corrupt and dictatorial regimes.

Although the mass demonstrations of the 1950s were a thing of the past, there was still great potential for ideological polarization. In particular, the revised US-Japan Security Treaty of 1960 called for 'prior consultation' between the two governments before putting into effect any major changes in the deployment or use of US forces stationed in Japan. There was particularly heated debate about whether this clause applied to the transfer of US forces and equipment from Japan to Vietnam. The use during the war of the bases in Okinawa, which was still under American rule, aggravated the controversy further.

Moreover, contacts with China — through non-governmental organizations (unions, opposition parties, intellectuals, etc.) and gradually expanding trade relations — had given the Japanese considerable insight into China's internal political changes and the difficulties in the Sino-Soviet relationship, and in this respect, too, they had their doubts about the efficacy of the US containment policy. As a result, even when Prime Minister Ikeda, with his essentially economistic stance, resigned on grounds of ill health and was succeeded by the ideologically more assertive and pro-US Sato Eisaku, Japan's cooperation in the war in Vietnam was no more than lukewarm. Prevented by its constitution from sending troops, Japan limited itself to providing economic assistance (about $55 million) and sending some medical teams to South Vietnam.[38] Its trade, however, did benefit from the Vietnam war (though to a lesser extent than it had from the Korean war), and Japanese exports to the participating countries rose considerably in the second half of the 1960s. It is estimated that the Vietnam war added over a billion US dollars to Japanese exports in 1971.

Thailand, for its part, comes somewhere between Japan and South Korea. Historically and ideologically, China and Vietnam represented the biggest threats to Thailand, so it was natural that it should have turned to the United States for support; the alliance with the United States was for a long time one of the basic tenets of Thai diplomacy. When, in 1954, John Foster Dulles, the US Secretary of State, proposed the establishment of SEATO, Thailand had been the first country to agree to join, although it felt some dissatisfaction that SEATO, unlike NATO, did not call for an all-out and automatic US commitment.[39]

At the time of the 1960–2 Laos civil war, the Sarit government had strongly urged American intervention on behalf of the anti-communist forces in Laos, but the Kennedy administration decided unilaterally not to commit its forces to that small, land-locked Asian nation and, instead, made a deal with the Russians at the 1961 Vienna summit to remove Laos from the agenda of contention between the superpowers. These developments taught the Thais the lesson that even questions which are of vital importance to a smaller ally may be of only peripheral importance to a superpower.[40]

When, however, American participation in the Vietnam war was at its height in 1965, the relative positions of the United States and Thailand altered. It was now the United States that needed Thailand's cooperation in order not only to make use of its airfields and logistic bases, but also to legitimize the US presence in Vietnam by giving the appearance, if not the substance, of sharing the burden with allies in Asia.[41] At least in the early stages, the Thais showed a marked reluctance to cooperate in the US war effort in Vietnam. Thus they insisted on 'prior consultation' for every planned use of their airfields, and not until 1967 did they officially admit that the bases were being used for bombing North Vietnam. Apparently they already felt some doubt about whether the United States had the will to continue the war to victory. Another factor, from 1964 onwards, was strong Chinese criticism of Thai-US collaboration as well as Chinese support for opposition movements inside Thailand.

While the most desirable position for Thailand would have been that of a spectator, able to observe from a safe distance the process of war, its geographical position and the existence of its air bases, indispensable to the US war effort, made an opportunistic attitude of this kind impossible.[42] So while the Thais tried as far as possible to avoid participating directly in the war, the Americans tried as far as possible to draw out their assistance and guarantees. Besides the routine assistance for specific items of cooperation, such as the despatch of Thai 'volunteer' regiments to Vietnam or the use of various military facilities in Thailand, the United States conducted an extensive programme of infrastructural development and periodical funding for counter-insurgency efforts.[43]

The scale of the American assistance, and the material benefits that it brought, had the effect of reinforcing the Thai military establishment's vested interest in the continuation of an American presence in the region; this, in turn, hampered the flexibility of Thai policies. Thailand also learnt that alliance with a superpower tends to breed ideological division domestically, and pushes dissident elements to seek support, overt or covert, from superpowers on the opposite side. The net result is that all the major powers are provided with pretexts to intervene by exploiting inherent divisions in the country, whether ethnic, religious or political.

Thailand was not the only country to absorb this lesson. Other governments in the region also learnt that in order to implement programmes of indigenous development, and thereby enhance their legitimacy — essential steps towards political stability — they must first set up the environment, individually or collectively, that would discourage outside powers from intervening in their internal affairs. It was this discovery that led in August 1967 to the establishment of ASEAN, the most notable example of regional cooperation, and to its subsequent success in developing a framework of operation. The need to preserve political independence and avoid becoming pawns in the superpower conflict is first enunciated in the organization's charter, the ASEAN Declaration, and is reiterated in successive declarations thereafter. The members of ASEAN, it is declared, must ensure freedom from 'external interference in any form or manifestation, in order to preserve their national identities in accordance with the ideals and aspirations of their peoples.'[44] Indeed, the events that followed were to prove that this kind of cooperation on the part of the smaller states in the region did much to protect the area from the 'domino' effect that the United States was so anxious to avoid.

Chapter Three
ASIA GOES MULTIPOLAR

Change in Indonesia

The abortive *coup d'état* which occurred in Jakarta in September 1965 was to have a profound effect on the pattern of political relations in Southeast Asia, accelerating a momentum of change which was already under way. Before the attempted coup — or Gestapu ('movement of 30 September') incident, as it is sometimes known — the Partai Komunis Indonesia (PKI) had been a vast organization, by far the largest party outside the communist bloc, with 3.5 million members. It enjoyed considerable political influence in Indonesia, as one of the pillars which sustained President Sukarno's 'guided democracy'. However, within a month of the mysterious events of the night of 30 September having taken place, the entire phenomenon — the party, the organization and its power — had disappeared completely. It was the greatest defeat the communists had ever encountered anywhere in the world and, as such, marked a turning-point in the history of the communist movement.

It was significant that neither China nor the Soviet Union was able to make even the slightest attempt to check a development of such magnitude. While a vast number of Indonesian communists and their sympathizers were being massacred, China remained utterly helpless, and the Soviet Union limited itself to blaming China for 'reckless adventurism'.[1] It was proof of the way in which Indonesia had managed to achieve freedom from great-power interference. Several years previously, the United States, too, was absolutely powerless when anti-communist and anti-Sukarno elements staged the so-called Permesta rebellion in Sumatra. If anything, the USA's feeble attempt to intervene on that occasion had merely pushed the country further to the left.

Admittedly, its geographical location made it difficult for either the United States or the Soviet Union, let alone China, to exert effective influence over the whole stretch of the Indonesian archipelago. More important, though, is the strong aversion that the Indonesians seem to have to any signs of interference, let alone domination, on the part of any of the major powers. This goes back to their revolutionary experience of 1945-9, when the United States was perceived as having obstructed the struggle for independence by taking sides with the Dutch, and the Soviet Union was regarded as having instigated an uprising against the young republic at Madiun in central Java.

Such experiences gave rise to deep mistrust of the great powers and helped to produce among both the elite and the masses a unique concept of 'free and active diplomacy' ('politik bebas dan aktif'), which was to become the guiding principle of Indonesia's foreign policy. This principle, reflecting a tolerance that is typical of Indonesian culture, is flexible — up to a point. In the words of a senior political leader, it is like a rubber plate, supple enough to be pulled either to the left or to the right as the situation dictates. However, it retains strong elasticity and whenever it is pulled too far in one direction, it bounces back strongly to the centre, where the true interests of the Indonesian people lie.[2] Even President Sukarno, the founding father of the Republic, could not break the rule with impunity. Towards the end of his career, when he seemed to have gone too far to the left, the 'rubber' bounced back, leaving him stranded.

For all his charisma and prestige as founder of the nation, Sukarno had been curiously alone throughout his life, having virtually no solid base of power among any of the major political forces in the country. In that, he was quite different from other leaders in Asia, such as Mao Zedong or Ho Chi Minh, who enjoyed overwhelming support and loyalty from broad sections of their party and the military.[3] Perhaps that was partly the reason why, towards the end of his career, he had to adopt the dangerous expedient of riding two horses, namely the army and the communists, a course which cost both him and the country dear. Having had to keep tension high all the time, he resorted repeatedly to brinkmanship politics, such as the territorial claims to West Irian or the confrontation (Konfrontasi) with Malaysia.

Sukarno had considerable skill in playing this game, particularly when it came to playing one superpower against the other. In the case of West Irian, his bluff paid off brilliantly in that President Kennedy had to acquiesce in his demand, but in the case of Konfrontasi he seems to have gone too far. In the middle of the night of

30 September 1965 an obscure commander of the president's guard ordered his troops to arrest a group of senior generals and hand them over to the mob, which murdered them at Halim military airport. Concurrently, the radio station was taken over, and the presidential palace was surrounded by the forces of a 'Revolutionary Council'. For some unknown reason, however, the dissidents spared General Suharto, then commander of the army's strategic reserve, who took control of the armed forces and pre-empted the full-scale uprising which the PKI was said to be staging. Sukarno's failure to disprove his own implication in the attempted coup critically undermined his authority. In March 1967, politically outmanoeuvred and effectively neutralized, he was obliged to relinquish office, and General Suharto took over the presidency.[4]

An interesting feature of this episode, revealed by subsequent events, is that the PKI was also strangely lacking in an independent base of mass support. In spite of its large-scale organization and enthusiastic following among broad segments of the peasant population, it somehow failed to turn these elements into revolutionary armies, or to secure territorial sanctuaries in the manner of the Chinese or Vietnamese communists. Towards the end, the party tried to build its own militia, on the pretext of fighting Malaysia, but it was too late to make any difference.

After the coup had failed, the communists were hunted down throughout the country by the army, which was quick to seize the opportunity to wipe out a rival which, under Sukarno's protection, had grown to be the most serious threat to its authority. While pledging to adhere to the 'free and active' principle in foreign relations, the Suharto government hastened to sever relations with China, on the grounds that communism, especially in its Chinese form, endangered Indonesia's domestic security.

The Birth of ASEAN

The departure of Sukarno symbolized the end of Southeast Asia's traumatic post-independence era and marked the beginning of a period when countries were pursuing the practical goals of national development. The shift which General Suharto began to bring about soon after the Gestapu incident met with unanimous approval from the neighbouring nations. Having been deeply disturbed by the empty but stinging rhetoric of Sukarno in such high-flown phrases as the 'Peking/Pyongyang/Phnom Penh/Hanoi/Jakarta axis', not to mention all the sabre-rattling of Konfrontasi politics, they were delighted to see signs of a 'good neighbour policy' coming out of this, the biggest nation in the area. Making the most of such a welcome trend, Thai Foreign Minister Thanat Khoman took the opportunity, at a banquet held in Bangkok to celebrate the end of Konfrontasi, to broach with his Indonesian counterpart, Adam Malik, the possibility of Indonesia joining an enlarged regional organization. Some such grouping could, he felt, replace the outdated ASA (Association of Southeast Asia), founded in 1961, which had been paralysed by various problems such as the Malaysian-Philippine dispute over Sabah, to say nothing of Konfrontasi.[5]

At this period the Thais already felt uneasy about the course of the war in Vietnam, which did not seem to be developing quite as the United States had planned. Also, they realized that the alliance with the United States held many hazards for the region, that it was unlikely to be permanent, and indeed that it was not desirable that it should be permanent. They were therefore searching for some indigenous arrangement which would represent the collective interests of the smaller nations in the region and, if possible, enable them to take a collective stand vis-à-vis countries outside the region, especially the superpowers. With Indonesia, the region's largest and internationally most prestigious country, apparently moving towards similar goals of stability and development, the Thais thought that a scheme of this kind at last stood some chance of being realized. In this sense, it can be said that ASEAN had its beginnings in the newly found congruence of interests of these two countries, Thailand and Indonesia.

The Suharto administration, for its part, realized that the proposed organization could offer an opportunity to fulfil the Indonesian people's longing to see their government play a leading international role, a role that was commensurate with the country's size, population and natural resource endowment, as well as its proud record of revolutionary achievement. The new administration was quick to see that ASEAN could provide, at least in part, a replacement for Sukarno's flamboyant and often capricious (though extremely popular) Afro-Asian diplomacy.[6]

Also, it might give the new government, which had inherited from Sukarno a shattered economy and a serious balance-of-payments problem, an opportunity to demonstrate that Indonesia was out to pursue a responsible and peaceful policy. By becoming a founder member of a forum in which Western-oriented countries such as Thailand and the Philippines were also participating, it would certainly be in a better position to ask for economic assistance from the developed countries of the West.

Nevertheless, in a still unsettled internal situation, the government had to be careful not to give the impression that it was moving too far to the right or abandoning the traditional principle of 'free and active diplomacy'. So Indonesia endeavoured to ensure that the proposed ASEAN would play down its Western bias (which all the prospective member states shared in one way or another) and project instead the image of being an independent, and indeed 'free and active', organization which aimed at the self-reliant pursuit of economic and social development. This gave ASEAN, right from the beginning, an emphasis on neutrality (though with anti-communist underpinnings) which was to gain greater significance in the later 1960s and the 1970s when the region went through many fundamental political changes.

This is not to say that ASEAN was without its problems in the 1966-7 period. Thailand, for example, was deeply involved in the Vietnam war as an ally of the United States and was hardly in a position to talk about neutrality, while the Philippines had no plans for the removal of the American bases on its territory. Singapore, which had British bases on its soil, did not want to accelerate the British withdrawal 'East of Suez' by any careless talk of 'neutrality'. All told, if a proposal for an ASEAN-style organization had been made in the first half of the 1960s, it is quite likely that problems of this kind would have caused ASEAN, like its predecessor ASA, to miscarry. By 1966-7, however, the regional and global environment was such that these countries no longer had much faith in their existing relations with the respective superpowers, and therefore they tried their best, under the given constraints, to come to terms with the proposed stance of ASEAN, which Indonesia had been instrumental in formulating.

For the Philippines it was a chance to play down its links with the United States, which it was beginning to find awkward to sustain, and to project instead an Asian image of the nation — a theme which was growing in popularity among Filipino liberals and the young. For Singapore, and also for Malaysia, it was a welcome chance to 'contain' Indonesia in a regional framework and thus to ensure its friendly attitude to the region. These two countries also faced the likelihood of a British Labour government deciding to pull out of its commitments 'East of Suez' earlier than scheduled, and were searching for some international forum which might replace the British connection, at least in part. For Singapore, the establishment of a regional forum which would ensure it a position of equality with its larger neighbours was a very welcome development. Ever since its independence, Singapore had had difficulty in establishing a proper identity in the region. The predominant Chinese-ness of its population was apt to provoke a variety of reactions from its big Malay neighbours, Malaysia and Indonesia, with which it was essential to maintain friendly relations; hence the persistent efforts on the part of the Singaporean government to project an image of a multiracial, multicultural and multilingual city-state. To participate in a new organization such as ASEAN, which would embrace nations of varying orientations, such as Thailand and the Philippines, would certainly help it to reinforce its regional identity.[7]

Thus the birth of ASEAN was in many ways a timely, and even a logical, development for the countries of the region, whose dominant concern at this time was to maintain their independence and integrity in the face of the open competition for influence among the major powers which they foresaw as following in the wake of the Vietnam war. Thanat Khoman describes the logic behind the formation of ASEAN most eloquently:

> As the political climate has changed from confrontation to peaceful competition and occasional negotiations among those endowed with greater power and resources, it becomes increasingly necessary for the small and weak nations to close their ranks and pool their limited means and potential. Nowadays, the risk of being swallowed up or trampled to death by the more powerful countries is so evident that the only viable course to follow is cooperation among those who share the same stake. As they used to say in a jocular vein, either we hang together or we hang separately.[8]

Thus ASEAN was duly launched at Bangkok on 7 August 1967, when the foreign ministers of the five member countries (Indonesia, Malaysia, the Philippines, Singapore and Thailand) signed a charter document, the ASEAN Declaration. On the role of foreign military bases, which was not only a sensitive point for each member state but also the most delicate issue in determining ASEAN's future orientation, the Declaration, reflecting the results of a complex process of multilateral compromise, states:

> All foreign bases are temporary and remain only with the expressed concurrence of the countries concerned and are not intended to be used directly or indirectly to subvert the national independence and freedom of states in the area or prejudice the orderly processes of their national development.[9]

Japan's Growing Involvement in the Region

Prime Minister Ikeda's emphatically 'economistic' stance paid off in the extraordinarily robust rate of economic growth which Japan began to show as it entered the 1960s. His aspiration to promote the nation's international status by means of its economic strength was fulfilled in 1964 when Japan was both accepted into the OECD and granted Article 8 status by the International Monetary Fund (IMF).*

In 1965 the government of Sato Eisaku, who had succeeded Ikeda as prime minister the previous year, achieved its first diplomatic success by opening up relations with South Korea in the shape of the Korea-Japan Treaty. The Sato government, which derived its support from right-of-centre factions in the LDP, was more sympathetic towards South Korea than the previous administration had been; strong opposition to the treaty came from the JSP, but the dissension did not reach anything like the proportions of the 1960 Security Treaty protests. In South Korea the strongly anti-Japanese president, Syngman Rhee, had been replaced by the more pragmatic Pak Chung-hee, who believed that Japan could help in his country's economic development. These changes on both sides enabled the debilitating legacies of mistrust and bitterness — dating back to the harsh Japanese colonial period in Korea — to be overcome sufficiently to allow the negotiations, which had been continuing intermittently for fourteen years, to reach a conclusion. Diplomatic relations were established, Japan pledged $500 million in economic assistance (in lieu of reparations), and South Korean fishing restrictions were lifted. But the territorial dispute over the island of Takeshima (Korean name, Dokto) had to be left for later settlement,

---

*Article 8 of the IMF's articles of agreement requires the removal of all restrictions on foreign exchange.

and the problem of the 600,000 ethnic Koreans inside Japan was not satisfactorily solved either. (Less than half that community was covered by the arrangement whereby those Koreans who registered as citizens of South Korea would be granted permanent residence.) Nevertheless, despite suspicions on both sides, economic and political links were to strengthen in the following few years.

In April 1966, at Japan's suggestion, a nine-nation Ministerial Conference for the Economic Development of South-East Asia (MEDSEA) was held in Tokyo. South Vietnam, Laos and Kampuchea joined the five future members of ASEAN in accepting the Japanese invitation to discuss prospects and policies for agricultural development and industrialization in the region. (Burma had also been invited, but refused.) Japan agreed to increase its economic aid to the region in what was seen as a sign of a new positiveness in its Asian policies.

On a broader regional (as opposed to sub-regional) basis, Japan became involved as one of the leading advocates of the Asian Development Bank (ADB), which began operation in December 1966. Japan and the United States were the two main donors. Although the headquarters were set up in Manila rather than Tokyo, the governors of the ADB have always been Japanese, and it is one of the few intergovernmental organizations where Japan maintains a high visibility.

Although the concept of an 'Asian Marshall Plan' had already been put forward in the early 1950s by Prime Minister Yoshida, it was really only from the 1960s that Japan began to be actively involved in Southeast Asia. Trade relations were initially stimulated by the reparations payments which Japan made to a number of Southeast Asian nations that had been victims of Japanese aggression in World War II. (Agreements were signed with Burma in 1954, with the Philippines in 1956, with Indonesia in 1958 and with South Vietnam in 1959, totalling $1.15 billion.) These payments usually took the form of capital goods, and this helped Japan to regain access to Asian markets and resources. With the country's economy growing so rapidly, the government had come to see the importance of Southeast Asia economically and politically, and began to take a much greater interest in the region's affairs. Thus Prime Minister Ikeda, after his visit to Jakarta in 1963, offered to mediate in the Malaysian-Indonesian confrontation, and, although this ultimately proved unsuccessful, a three-country summit meeting — President Sukarno from Indonesia, Prime Minister Tunku Abdul Rahman from Malaysia and President Diosdado Macapagal from the Philippines — was held in Tokyo in June 1964. The meeting ended without agreement, but Japan continued to take an interest in the dispute, and during 1965-6, by offering conditional aid, was able to moderate Indonesia's militant stance towards Malaysia and reduce its dependence on China. This can perhaps be described as the first post-war political initiative that Japan had made in the region.[10]

Japan looked upon Indonesia, the largest Southeast Asian nation and an invaluable source of raw materials, as the most important in the region. Although there had been some problems, such as alleged misuse of funds and the mismanagement of projects based upon reparations payments, the economic relationship between the two countries had become increasingly close. Consequently, Japan had viewed the abortive coup of September 1965 and ensuing events with considerable concern, and had welcomed — subsequently — the rise of the Suharto government, with its publicly stated aim of political stability through economic and social development. Therefore, when the new Indonesian government was forced to accept that international assistance was necessary for the rebuilding of its economy, Japan, with some prodding from the United States, agreed to become one of the main contributors to the international aid consortium, known as the Inter-Governmental Group on Indonesia (IGGI), which was established in February 1967.

Japan had begun to give aid, though in modest amounts, in the 1960s, with special emphasis on assisting the development of the countries of the Asian Pacific region. In 1963, 56 per cent of the total flow, including loans and investment, went to Asia, and 99 per cent of the total Official Development Assistance (ODA) was directed towards Asia, with Southeast Asia claiming half of this amount. Official Japanese figures for the cumulative totals for Japanese economic assistance up to 1975 show that 78 per cent of this went to six East Asian countries, namely South Korea, Burma, Malaysia, Indonesia, the Philippines and Thailand.

Japan's investment in East Asia was still relatively insignificant in the first half of the 1960s, but in the second half of the decade there was an increasing flow into Hong Kong, Singapore and, above all, Taiwan, which after the completion of the Kaohsiung Export Processing Zone in 1966 became the principal East Asian recipient — at least until the early 1970s. These countries attracted, in particular, Japanese manufacturers of labour-intensive products; third-country markets were the targets as much as local markets. During this period the East Asian countries began to pursue an 'outward-looking' strategy of economic development which called for the encouragement of inflows of foreign capital and technology to build up export industries. However, even with the establishment of diplomatic relations with South Korea in 1965, it took another two years before the first Japanese investment was authorized, and levels only really took off in the early 1970s. Up to then, one of the main characteristics of Japanese investment in the above four countries was — and to a certain extent this is true even now — the small average scale of the enterprises involved. In the case of Malaysia, Indonesia and the Philippines, by contrast, Japanese investment, though quite small in total until the beginning of the 1970s, consisted mainly of larger-scale resource-extraction operations. Thailand, the odd man out of the eight East Asian countries, adopted very liberal foreign investment policies early in the 1960s and therefore was already attracting considerable Japanese capital in that decade.

It was in trade that Japan's impact was greatest in the 1960s: the total value of its trade with the five ASEAN countries as a whole nearly quadrupled between 1960 and 1970, and that with the three newly industrializing countries (NICs) of Northeast Asia grew even faster — in South Korea's case by nine times. Japan's exports to these eight countries grew from $838 million in 1960 to $4 billion in 1970, and its imports from $686 million in 1960 to $2.4 billion in 1970. However, apart from Indonesia and Malaysia, the other six East Asian countries suffered trade imbalances with Japan for most of the decade. South Korea and Taiwan excepted, all the East Asian nations found their dependence on Japan — in terms of its share of their imports and exports — increasing (see Table 3.1).

Table 3.1: Japan's Share in the Exports and Imports of the other East Asian countries (%)

		1955	1960	1965	1970
Hong Kong	exports	5.8	5.8	5.9	7.1
	imports	14.1	16.1	17.2	23.8
South Korea	exports	52.6	62.5	26.0	28.0
	imports	15.7	20.2	37.1	40.8
Taiwan	exports	59.3	37.6	31.1	15.1
	imports	30.3	35.5	40.2	42.8
Indonesia	exports	7.4	4.1	19.8	40.8
	imports	14.4	16.0	37.7	29.4
Malaysia	exports	7.2	15.7	24.0	18.3
	imports	5.8	7.1	16.2	17.5

Table 3.1: — continued

		1955	1960	1965	1970
Philippines	exports	15.2	24.0	28.3	40.1
	imports	7.8	23.3	23.8	30.6
Singapore	exports	n.a.	4.5	n.a.	7.6
	imports	n.a.	7.3	n.a.	19.3
Thailand	exports	17.4	18.1	18.8	25.5
	imports	21.1	26.2	31.6	37.4

Note: The 1955 and 1965 figures for Malaysia are proportions of the combined figures of Malaysia and Singapore.
Source: Andras Hernadi, *Japan and the Pacific Region* (Hungarian Scientific Council for World Economy, Budapest, 1982).

It was in the 1970s that Japan began to buy significant quantities of Indonesian oil and timber and Malaysian tin and rubber, but the other countries found it more difficult to sell large enough quantities of their products to Japan to counterbalance their growing imports from Japan. For all eight countries the Japanese share in their imports rose during the 1960s — and in the case of Malaysia, Singapore and South Korea more than doubled — with that share ranging from 18 per cent in Malaysia to 42 per cent in Taiwan by 1970. The Japanese share of the five ASEAN countries' exports also increased during the decade, but decreased for South Korea and Taiwan and remained almost static for Hong Kong. In the early 1970s the continuing trade imbalance, the growing dependence on Japan for markets and supplies, and the more visible Japanese investment were to lead to criticism from Southeast Asian countries, in particular Thailand, of 'over-presence' and domination by Japan.

As for ASEAN, Japan did not show much excitement, at least in the early stages of its existence. By rights, ASEAN, as a vehicle for regional cooperation, should have been welcomed by Japan, but it was mistakenly taken to be yet another anti-communist alliance, along the lines of the Asian and Pacific Council (ASPAC), which was formed in June 1966 by nine East Asian and Australasian countries, primarily at South Korea's instigation. Some people even suspected that it might turn out to be a 'pressure group of primary product producers knocking on Japan's door'.[11] Even at the governmental level, there was little interest. For instance, in a policy report which Prime Minister Sato presented to the Diet in December 1967, right after his two long trips to Southeast Asia and Australia in September and October 1967, there was not one word about ASEAN.

Japan's response is a measure of the low international esteem in which ASEAN was at first held. Obviously, given the situation of the world under near-total domination by the superpowers, the formation of a political association consisting of a handful of developing countries did not carry much weight. Few people inside or outside the region had any thought that such a tiny organization could ever become an effective regional counterbalance to the juggernaut of superpower politics. As if to prove the point, ASEAN itself was beset with problems in its early days. With Indonesia near to bankruptcy and the other countries not much better placed, it could clearly be a very long time before the development process would really take off. Regional disputes, including the Sabah question between Malaysia and the Philippines, were still unresolved, and insurgencies were continuing in one way or another in most of the countries. The diverse diplomatic stances of the five countries suggested that a cohesive stand on external policies would be difficult, if not impossible. Indeed, it took several years and a drastic change in the international environment itself before ASEAN was internationally recognized as an effective instrument for regional cooperation, with a key role to play in the stability and development of the region.

Nevertheless, ASEAN's avowed aim of building 'peace, freedom and prosperity' through its own efforts was highly compatible with Japan's emerging goals. Apart from geographical proximity and feelings of cultural affinity, Southeast Asia was important to Japan as a source of economic security, providing resources, markets and investment sites as well as maritime communications. In order to maintain that security, a prerequisite for Japan was, first and foremost, the political stability of the region, and this ASEAN was able to bring about in the years to come.

Détente and Balance-of-Power Politics

On 30 March 1968, US President Lyndon Johnson, in a nationwide broadcast, announced a halt to the bombing of North Vietnam and proposed that discussions on the possibility of peace talks should take place in Paris. This was a *de facto* admission of defeat in the war, and, as a way of accepting responsibility for that, he announced that he would not be standing in the presidential election set for November that year. Without doubt this had been an agonizing experience for Johnson, under whom the Vietnam war had gradually escalated, from 16,500 military advisers in the Kennedy era to the use of 500,000 ground troops, with an annual expenditure of $25 billion and the broadening of the war to all areas of South Vietnam. However strong the American economy was, it clearly could not tolerate expenditures and sacrifices on this scale indefinitely. For one thing, these costs were a great obstacle to the construction of the 'great society' which was one of the central goals of the Johnson administration.[12]

US field commanders in South Vietnam had been advising that in order to win the war there was no other course but to invade North Vietnam, for which purpose 700,000 troops would be needed. However, the February 1968 Tet offensive, in demonstrating the extraordinary capability and morale of the National Liberation Front, upset American calculations, and added fuel to anti-war sentiment inside the United States which, reinforced by mass media coverage of the grisly realities of the war, was rapidly reaching fever pitch.[13]

As though trying to console himself, and his fellow Americans, for the nation's failure, President Johnson referred in his speech, almost diffidently, to the latest developments in Indonesia and other parts of Southeast Asia as evidence of a positive result that the war may have had:

> Since 1966, Indonesia, the fifth largest nation in the world, has had a government dedicated to peace with its neighbours and improved conditions for its own people. Political and economic cooperation between nations has grown rapidly . . . Every American can take pride in the role we have played in Southeast Asia. We can rightly judge — as responsible Southeast Asians themselves do — that the progress of the past three years would have been far less likely, if not impossible, if America and others had not made the stand in Vietnam.[14]

It is probably stretching the point too far to take the Gestapu incident in Indonesia in September 1965 as resulting from the Vietnam war, for this episode, if it can be explained at all, must be attributed to the internal dynamism of Indonesia itself. Nevertheless, the new realism emerging in Asia, the emphasis on self-reliance, and the embryonic but nevertheless conspicuous trend towards regional cooperation, together with the continuing economic growth of Japan and those countries which later came to be known as the Asian NICs, all contributed to the re-evaluation that the United States now undertook of its Asian policy, which, for twenty long years, had centred on the aim of 'containing China'.

* * *

The new administration of Richard Nixon, sworn in in January 1969, was fully conscious of the fact that the basic concept which had sustained US foreign policy for the past two decades, and which had led to the war in Vietnam, was no longer relevant in the world of the late 1960s. A viable alternative had to be found if the new administration was to regain the initiative in foreign policy matters and effectively overcome the despair and frustration of the American public. Dr Henry Kissinger, newly appointed as the President's Special Adviser on National Security, had written:

> The United States is no longer in a position to operate programs globally; it has to encourage them . . . In the forties and fifties, we offered remedies; in the late sixties and in the seventies our role will have to be to contribute to a structure that will foster the initiative of others. We are a superpower physically, but our designs can be meaningful only if they generate willing cooperation. We can continue to contribute to defense and positive programs, but we must seek to encourage and not stifle a sense of local responsibility.[15]

Earlier — in 1967 — Nixon, too, had attempted to articulate a new line of thinking on Asia in a similar context. Recognizing that for the United States to 'go it alone' in containing China would impose 'an unconscionable burden' on itself as well as undermine the development of the Asian nations, he urged that the 'primary restraint on China's Asian ambitions' should be 'exercised by the Asian nations in the path of those ambitions, backed by the ultimate power of the United States'.[16]

The culmination of this rethinking was the new policy orientation outlined by the President at a press interview in Guam on 25 July 1969. The wording, as reported in the *New York Times*, is of some interest. Summarizing what has come to be called the Guam Doctrine, Nixon said that the fastest rates of growth in the world were occurring in non-communist Asia — namely in Japan, South Korea, Taiwan, Singapore and Malaysia — and that 'regional pride' and a sense of 'Asia for the Asians' were becoming major factors. The

United States would continue to be a Pacific power, but that:

> As far as the problems of international security are concerned, as far as the problems of military defense, except for the threat of a major power involving nuclear weapons, [it] was going to encourage and had a right to expect that this problem would be increasingly handled by, and the responsibility for it taken by, the Asian nations themselves.

He also said that the United States would not undertake 'the kind of policy that will make countries in Asia so dependent upon us that we are dragged into conflicts such as the one we have in Vietnam.'[17]

This new orientation meant a clear departure from a policy framework which had determined the political landscape of East Asia for two long decades. It is not surprising that it had an unsettling effect. Thailand, for instance, had already been thoroughly alarmed by the speech in which President Johnson offered North Vietnam the possibility of settlement through negotiation. At that time, the United States had 50,000 men stationed in Thailand, and Thailand itself had committed over 12,000 men in Vietnam. In the event of the United States opting to acquiesce in North Vietnam's eventual domination over the whole of Indochina, Thailand, which, for all its hesitancy, had long cooperated in the American war effort, would be forced to make very difficult and possibly quite costly adjustments. It was doubly painful for Thailand that such an important decision had been made unilaterally by the United States, without prior consultation with it. Now, after Nixon's speech, the Thais were faced with the daunting task of having to repair the framework of their diplomacy, which had been ravaged by their overcommitment to the United States, and to recover their traditional freedom of manoeuvre.[18]

The impact of so great a shift in American policy was not confined to its allies. The Russians saw in the Guam Doctrine a danger that the United States would give too much freedom and initiative to its Asian friends. They thought that the United States was now trying to forge a new type of alliance consisting of the pro-US nations in Asia, whereby China would be controlled by a multiple balance of power in the region rather than by direct containment by force, a course which was no longer feasible. Such a development was not altogether to their liking. For, although enhanced regionalism might liberate the United States from certain responsibilities, it would in turn encourage regional actors to manage their own affairs independently, which was not necessarily in the Russians' interest. They appear to have felt more comfortable in a bipolar system, in which they had to compete only with the United States. Also, understandably, they were loath to see China being prodded by the United States out of its diplomatic isolation and allowed to build up independent relations with various actors in the region on its own initiative.[19]

China, too, seems to have seen a danger signal in the Guam Doctrine in that it would liberate Japan to pursue its traditional 'imperialistic' purposes in the region, this time with the express understanding and support of the United States. Having 'woken up' from the trauma of the Cultural Revolution, which had engaged all its attention internally, China discovered, perhaps to its surprise and dismay, that Japan had grown into a substantial economic power, having world-wide commercial links and considerable influence over the economy of Southeast Asia. Joint ventures were springing up everywhere, and a whole range of consumer goods were flooding the region, which was having an undesirable influence on the masses of Southeast Asia, who were to have been the next target for China's ideological expansion. Also, as Zhou Enlai stated later (in August 1971), the Chinese were genuinely concerned about the possibility of Japan reasserting its colonial domination over Taiwan and South Korea, under the aegis of the United States. This may have been one of the reasons why China decided to accelerate the process of rapprochement with the United States and establish a direct channel of dialogue with the Americans which would give it leverage in the future management of the Asian Pacific region, including a chance to check any undesirable advance by Japan.[20]

The fear and anger which China had felt towards the Soviet Union when it invaded Czechoslovakia in August 1968 would have further encouraged a policy reappraisal of this kind. The 'Brezhnev Doctrine', as subsequently expounded, must have been anathema to the Chinese in that it gave the Russians virtually a free hand to intervene in the internal affairs of its allies-cum-satellites, by force if necessary, whenever their internal situations were regarded as jeopardizing the progress of the world socialist movement.

Under these circumstances, it was only a matter of time before a Sino-US rapprochement took place. Only two weeks after Nixon's inauguration, a presidential directive set in motion the discreet and complex process of establishing a channel of communication with China, culminating, on 15 July 1971, in the President's surprise announcement on television of the confidential trip made by Kissinger to China and the agreement which he had reached with the Chinese about a subsequent visit by the President himself. It was a worthy result for an administration which looked for spectacular political gains in the field of foreign relations. In a single dramatic step, it altered the age-old framework of relations. It sent a shock wave through the nations in the region, both friends and adversaries, who were now forced to adjust their policies, often with considerable pain.

The North Vietnamese were angered by what the armed forces' journal, *Quandoi Nan Dhan*, described as 'Nixon's plan . . . to divide the sphere of influence between the big powers and split the socialist countries and people's liberation movements',[21] and by the prospect of the big powers colluding to determine the fate of their smaller allies *in absentia*. Appreciating this fear, Zhou asked a visiting French parliamentary group, which was on its way to North Vietnam in February 1972, to explain to their hosts in Hanoi that the question of Vietnam would not be on the agenda of the impending talks between him and the President. Nevertheless, after Nixon's visit to China, Kissinger revealed to the press that Vietnam had indeed been discussed in the talks in Beijing, and that a reduction of the US military presence in Taiwan was also mentioned as one means of reducing tension in the region.[22]

Dilemma in the Korean Peninsula

The change in US global policy and the subsequent Sino-US rapprochement had a particularly profound impact upon the nations of Northeast Asia. Obviously the hardest hit was Taiwan. The rise of mainland China's position in the international community spelled a corresponding decline in the role of Taiwan. Not only was its status in the international community undermined, but the network of relations which it had painstakingly built up over many years was devastated. As for Japan, where the term 'Nixon shokku (shock)' was coined, America's sudden about-face shook the nation to the core. Particularly hurtful was the fact that the Nixon administration did not warn Japan of the new development in its China policy until literally a few minutes before the famous TV announcement. (The 'Nixon shock' is discussed in detail in the next chapter.)

But it was North and South Korea for which the future held the greatest uncertainty. Their nationhood had been built on the premise of the cold war, and the multipolarization of Asia was bound to undermine the foundations of their political structure. At the very least, they were frightened by the likelihood that the fate of the Korean peninsula could now be discussed by the leaders of China and the United States without either of the Koreas being represented — a state of affairs which, clearly, they found totally unacceptable.

On 12 August 1971, less than a month after Nixon's announcement of his overture to China, the South Korean Red Cross proposed discussions about allowing visits by divided families across the border. The North Korean Red Cross promptly accepted, and on 20 September there began the first of what were to become weekly meetings between the Red Cross representatives of both sides. The eagerness of both countries to launch some kind of a dialogue demonstrates the unease which they felt about the emerging new political structure in the region.

On 4 July 1972 the two countries issued a joint communiqué which stipulated that reunification should be achieved through independent Korean efforts, without outside interference, and by 'peaceful means'. It also provided for the establishment of a North-South Coordinating Committee, which was to meet several times during the following year until its suspension in August 1973. From the South, Lee Hu-rak, director of the Korean Central Intelligence Agency (KCIA), and from the North, Kim Yong-ju, brother of President Kim Il-sung, were appointed co-chairmen of the committee. However, despite this display of enthusiasm for dialogue, little progress was made in substance, showing that the entire exercise was intended primarily to demonstrate the two countries' ability and intention to determine their fate for themselves, and that, apart from that, there was little common ground between them.[23]

The changing pattern of political relationships in the region presented both countries with complex problems in different ways and yet in somewhat similar contexts. Understandably, the North Koreans were frightened by the new dimension that the Sino-US rapprochement introduced into the Sino-Soviet dispute, which had long been their most difficult problem to deal with. Since the beginning of the rift between the two communist giants, the allegiance of North Korea had oscillated between the two. When Khrushchev launched his criticism of Stalin in 1956, North Korea had leant towards the Chinese side, since it found the new Soviet 'revisionism' unsettling, while the 'pure' revolutionary rhetoric of China was more in harmony with its own ideological orientation. (For example, in some respects, China's 'Great Leap Forward' was paralleled by North Korea's 'Flying Horse Movement'.)

However, it was also an unsettling experience to be at odds with the Soviet Union, so that, after the fall of Khrushchev in 1964, North Korea worked at improving ties with the Soviet Union, which responded with a visit to Pyongyang by Premier Kosygin in 1965. Then, in 1969, when it saw a groundswell of change developing in international relations in Asia — typified by the Guam Doctrine — North Korea again moved back towards China, frightened perhaps by the possibility of China leaning towards the United States and, by implication, towards South Korea, and was rewarded by a visit from Premier Zhou Enlai to Pyongyang in 1970.

These manoeuvres, however, could give North Korea little solace. In the face of the widening schism between China and the Soviet Union, now symbolized by the Sino-US rapprochement, not even the concept of shared purpose among communist nations, including such a fundamental premise as the common struggle against imperialism, could be taken for granted. This could obviously spell very grave danger for North Korea, located, as it was, on the periphery of both China and the Soviet Union and at the same time confronting the might of the 'imperialist' forces across the 38th parallel. In these circumstances, it was only natural that it should try to become more self-reliant and thereby to reduce its dependence on outside powers, by however little. It therefore concentrated, internally, on accelerating economic development and, externally, on widening its relations, as seen in its admission during 1973 to three of the major international organizations: the Inter-Parliamentary Union (IPU), the World Health Organization (WHO) and the United Nations Conference on Trade and Development (UNCTAD).[24]

For South Korea, the first ominous effect of the Guam Doctrine was America's decision, taken unilaterally in March 1970, to withdraw one division of its troops. It was a blow to the Pak government, which took it as a betrayal of the trust and friendship that had been carefully nurtured between the two nations ever since the Korean war. This sense of betrayal was heightened by the fact that the Koreans had believed that, having cooperated with the United States in its war in Vietnam by sending troops there at substantial cost and sacrifice, they had established some sort of bargaining power vis-à-vis the United States. They assumed therefore that the United States would at least consult with them when it decided on a policy which might be detrimental to their interests.

In April 1971, President Pak received a considerable blow in the presidential election, in which, contrary to his expectation of an easy victory, the opposition candidate, Kim Dae-jung, came within less than a million votes of defeating him. After the election, however, Kim was in turn defeated in an intra-party struggle and left for self-imposed exile in the United States and Japan, purportedly to carry out his anti-Pak campaign outside Korea. Although this temporarily removed an irritant for the government, Kim's subsequent fate — kidnapped in Japan in June 1973, allegedly by the KCIA — continued to threaten Pak's position throughout the 1970s.

On top of all this, Nixon's announcement in July of the Sino-US rapprochement had an even more shattering effect on Pak and his government than it had had on the North Koreans. The implication that, as far as East Asia was concerned, the cold war — which provided the foundation of South Korea's security — was losing momentum could only mean an unpredictable future for the region. Spurred on by a great sense of insecurity, South Korea hastened to initiate the dialogue with the North — as an emergency measure to protect its minimum basic interests, rather than as a well-thought-out policy. As in the case of North Korea, the South developed a sort of siege mentality, which magnified, perhaps out of proportion, the need to close ranks and promote national self-reliance, economically, socially and politically, as the only way to reduce its dependence upon outside powers whose reliability now looked so dubious.

On 17 October 1972 President Pak imposed martial law, dissolving the National Assembly and revoking the Constitution. On 22 November a national referendum gave results showing that 98 per cent of the population supported the proposed new 'Yushin Constitution' (Renovation Constitution), which restricted popular rights and greatly strengthened the authority of the President. Thus, to start with, the dictatorial system that was to dominate South Korea in the 1970s had a fair measure of support, reflecting, it seems, the general agreement of the public with President Pak's contention that it was essential to eliminate, even if only temporarily, all opposition and political inefficiencies if he were to carry out the delicate process of dialogue with the North as well as to safeguard the interests of the South in an increasingly unpredictable international situation.

Indeed, although welcomed by many as the first step towards reunification, the joint communiqué of 4 July 1972 provoked strong opposition, particularly in military circles and in the refugee groups (from the North), both of which had considerable political influence in the country. They argued that dialogue, taking place in the partial vacuum created by US troop withdrawals, could only lead to a further weakening of the country's defence system. It was not constructive opposition, since they had no alternative to offer to the dialogue that was already under way, but nevertheless the government had to heed it. President Pak countered by claiming that, in order to engage in the negotiations with the monolithic North, the South must be equally united behind the government. Also, there seems to have been a consensus that, to judge by the experiences of the 1960s, Western-style democracy was not wholly compatible with the political culture of Korea, and that now might be the time to experiment with something new.[25]

Despite considerable effort on both sides, the North-South dialogue did not proceed as hoped. The North Koreans became increasingly obstructionist, as shown in their insistence — in talks held in November 1972 — that the negotiations should take as their starting-point such difficult topics as a halt to the military build-up, the withdrawal of US troops and interchange between political parties, whereas the South Koreans urged building up an interchange in those areas which at least had some possibility of being realized, such as economic and cultural exchange. As a result, the dialogue soon reached deadlock.

Perhaps North Korea began to realize that one of its principal aims in entering into dialogue, namely to accelerate the process of US troop withdrawal from the South by advocating détente, was not going to be accomplished so easily. Moreover, the holding of meetings alternately in Pyongyang and Seoul meant that many North Koreans would visit the South, which might not be particularly desirable. It could even be dangerous to allow North Koreans to see the reality of the material prosperity of pre-oil-shock Seoul, which was widely at variance with the version put out by North Korean propaganda: severe exploitation of the workers, the perennial poverty of the proletariat and a society in imminent danger of collapse.

As time went on, North Korea intensified its obstructionist tactics, as illustrated by the totally unrealistic idea of the creation of a 'Confederate Republic of Koryo' which it put forward in June 1973 in response to the South's proposal for joint entry into the United Nations. Finally, on 28 August 1973, using as a pretext the Kim Dae-jung abduction incident (which had occurred in Tokyo that June), Kim Yong-ju, the North Korean co-chairman of the North-South Coordinating Committee, openly criticized his South Korean counterpart, Lee Hu-rak, and unilaterally declared the suspension of further dialogue. Thus, after two years, the talks broke down.²⁶

This state of affairs was in clear contrast with the situation in Europe, where an easing of tensions produced the Ostpolitik. In Asia, too, the détente resulted in many new situations, such as Japan's rapprochement with China and the normalization of its diplomatic relations with Vietnam. That the same did not happen in Korea is due perhaps to the fact that in this case the ideological conflict was so closely interwoven with the conflict between North and South.

The failure of the North-South talks, although half expected, produced particularly unfortunate results for the South, because it took away the government's sole justification for applying the restrictions embodied in the Yushin Constitution. The termination of the dialogue inevitably released the people's pent-up resentment of the government's autocratic behaviour. Student demonstrations, demanding the truth about the Kim Dae-jung affair and an end to intellectual fascism as well as to subservience to Japan, took place at Seoul National University in early October and were echoed at other universities, so that during November most universities had to be closed down.

In December 1973, President Pak removed Lee Hu-rak from the directorship of the KCIA in what might have been intended as a conciliatory gesture on the part of the government. The opposition, however, which now embraced not only students, but also a wide range of the general population, including Christians and intellectuals, chose to interpret this as a sign of weakness; their activities grew more strident and they organized a petition with one million signatures demanding the revision of the Yushin Constitution. No longer able to cite the North-South dialogue as an excuse for its repressive measures, nor able to give in to public pressure, the only course left to the government was to intensify its autocratic stance. In January 1974, new emergency measures were introduced under which anyone who discussed or proposed revising the Constitution could be sentenced to up to fifteen years' imprisonment; and in April all student political activities were prohibited on possible penalty of death.²⁷

North Korea did not fare much better. Operating under somewhat similar conditions of threat and insecurity, the Kim Il-sung government launched in 1973 a crash programme of economic development designed to accelerate the completion of the existing six-year economic programme by a full two years. This meant a drastic reshuffle of the cabinet and party hierarchy, and a mobilization of the people on a scale that was comparable only with that carried out in China under the Cultural Revolution. It is ironic that owing to the process of détente and multipolarization in Asia, North and South Korea had to go through somewhat similar processes of adjustment, both of which were to leave deep scars on their respective political structures in the 1970s.

A Zone of Peace, Freedom and Neutrality (ZOPFAN)

In Southeast Asia, an additional impetus was given to the process of multipolarization by Britain's decision to withdraw its armed forces 'East of Suez' before the middle of the 1970s. In the old bipolar world, the British presence had been beneficial to the two regional members of the Commonwealth, Singapore and Malaysia, both in shielding them from excessive involvement in the US war effort in Vietnam and in acting as a deterrent against regional threats such as the one from Indonesia during the Konfrontasi. Moreover, since Britain was only a junior partner in the Western security system, there was little danger of the British umbrella implicating its clients in a global confrontation. Lee Kuan Yew was therefore strongly against the British withdrawal. When the Labour government brought forward the proposed schedule of withdrawal to March 1971, instead of the originally planned mid-1970s, he immediately flew to London to make a last-minute appeal — to little avail, since the withdrawal programme was a part of the Labour Party's declared policy.[28]

Malaysia's reaction to the British move was more complex because its domestic politics were sharply polarized along the ethnic lines of its population. The ethnic Malay leadership had always been sympathetic to the concept of non-alignment, following the example of Indonesia as well as of Islamic countries in the Middle East; this had precluded the government from taking a pro-US stance (e.g. by participating in SEATO), which would have produced a violent reaction from Indonesia, and in turn would have fermented opposition among the Malay population. Even on relations with Britain, the government had to tread carefully, because there was a strong undercurrent of opposition among the Malays, notably among such figures as Dr Mahathir Muhammad (later Prime Minister), who criticized Tunku Abdul Rahman, the then prime minister, for allowing Britain to exert influence on Malaysian foreign policy as the price for the Anglo-Malaysian Defence Agreement.

Therefore, the Malaysian response to the new time-limit for the British presence was to explore means of providing a new arrangement to replace it, rather than to try to prolong it. The result of this attitude on the part of the Malays was the Five-Power Defence Arrangement, which was signed in November 1971, by Britain, Australia, New Zealand, Singapore and Malaysia. It was clear, however, that the real advantage of this arrangement lay in the psychological reassurance that it gave rather than in any substantive commitment. It could certainly not meet the wide range of Malaysia's security needs in the unpredictable post-Vietnam period.

Against this background, Malaysia, in an attempt to forge a foreign policy which would satisfy the real aspirations of its people, conceived, and tried for some time to popularize, a revolutionary concept which called for the neutralization of Southeast Asia, with an agreement from the great powers that they would exclude the region from their own power struggles. When this idea was first broached by Tun Ismail, a senior Malay politician, in 1968, it received spontaneous support from all major groups in Malaysia. The ethnic Malay population welcomed it because it meant a move away from a pro-West position towards a non-alignment stance, and the Chinese liked it because it implied a fairer approach to China than before.[29]

This truly national consensus enabled the Malaysian government to launch the idea regionally — an opportunity it had been looking for, since, in the aftermath of the race riots in Kuala Lumpur in May 1969, the nation obviously needed to restore its image by making a bold policy move. Therefore, when Tun Razak succeeded the Tunku as prime minister in 1970, he immediately took it up as one of his diplomatic priorities. At the Fourth Ministerial Meeting of ASEAN, held in Manila in March 1971, Tun Ismail explained the concept as follows:

> It is with Vietnam in mind, together with the withdrawal of the American and British from Southeast Asia, that our government is advocating a policy of neutralization of Southeast Asia to be guaranteed by the big powers, viz. the United States, the Soviet Union and the People's Republic of China. The policy is meant to be a proclamation that this region of ours is no longer to be regarded as an area to be divided into spheres of influence of the big powers.[30]

The other ASEAN countries agreed in principle to the basic premise of the Malaysian proposal. In view of its long history of domination by outside powers, whether Western colonialism, Japanese militarism or global ideological struggles, it was time for Southeast Asia to call an end to external interference and start the work of national — and regional — development. Moreover, the ASEAN countries were quick to realize that a policy of this kind was consistent with the current trends of détente and multipolarization in global politics. Nevertheless, some of them also had reservations.

The Thais were not wholly at ease with the word 'neutrality', because of their long experience of fighting against the 'neutralization' of Laos as well as their long-standing animosity towards the 'neutral' stand of Cambodian Prince Sihanouk, while the Indonesians would have felt happier if they had been the ones to make the proposal. In view of their distinct neutralist orientation, deeply rooted in their traditional and near-sacred principle of 'free and active diplomacy', they felt that they were better qualified to make such a move than the Malaysians, whose record on 'neutrality' was not all that clean. Ever since the Sukarno days, the Indonesians had held a vision of an integrated Southeast Asia constituting a bulwark against external influence and intervention, and had refused to accept the need for any outside powers to fill the so-called vacuum created by the Western retreat.[31] In addition, they were a little apprehensive about the provision of the 'guarantee' to be obtained from the major powers, since it might force them into opening relations with China.[32]

All the ASEAN member countries, however, recognized the importance of the proposal for the new Malaysian government in its effort to consolidate its position vis-à-vis the various groups inside the country. By this time, it was generally accepted by the member governments that one of their most important collective functions was to help other member governments to reinforce their internal positions. In the four years since ASEAN's inception in 1967, they had come to learn that certain regional actions, such as hosting ASEAN Ministerial Meetings or having proposals accepted as ASEAN policy, were extremely useful in enhancing the credibility of governments in their own countries, and they were willing to extend such assistance to each other whenever possible. Consequently, at the Fifth Ministerial Meeting, held in Kuala Lumpur in November 1971, the five nations issued a joint declaration, endorsing Malaysia's proposal on the neutralization of Southeast Asia as a 'desirable' objective and stating that:

Indonesia, Malaysia, the Philippines, Singapore and Thailand are determined to exert initially necessary efforts to secure the recognition of, and respect for, Southeast Asia as a Zone of Peace, Freedom and Neutrality, free from any form or manner of interference by outside power.[33]

At the outset, the Kuala Lumpur Declaration (or ZOPFAN Declaration as it is sometimes called) was not taken seriously by many outsiders, who dismissed it as a 'maiden's prayer', meaning that it was no more than wishful thinking. A somewhat cynical view prevailed, generated by the cold war, that, whatever the small countries wished or decided, when the chips were down their fate would be at the mercy of the big powers. The conspicuous lack of practical achievement in the first four years of ASEAN's existence did little to suggest that this time, thanks to a new collective will, things would be different.

At the very least, two things were needed if the declaration was to have substance. One, as Tan Sri Ghazali Shafie, then Permanent Secretary at the Malaysian Foreign Ministry, pointed out, was for ASEAN to 'devise ways and means of, and undertake responsibility for, ensuring peace among member states'. The other was to promote national and regional self-reliance, economically and socially, so that 'a neutralized Southeast Asia would meet the basic legitimate interests of the great powers themselves', as Tun Razak asserted in a speech in July 1971. Foreign Minister Carlos Romulo of the Philippines commented that it would require a transitional period, a time of experiment, before final commitment was made to the ZOPFAN concept, because the countries would have to re-examine traditional alliances and revise long-standing arrangements. Both the Philippines and Thailand were bound by an alliance relationship with the United States. Also, as the Foreign Minister of Singapore later commented, as long as the war was still being fought in Vietnam, it was difficult to ascertain what kind of political structure would emerge at the end of it, or how the easier atmosphere between the superpowers would actually affect Southeast Asia.[34]

Although ASEAN deployed the concept of neutrality in the Kuala Lumpur Declaration, this was really an endeavour on its part to ensure freedom of action in the event of political change in the region. In view of the fact that a majority of the developing nations of the world — in the Middle East, Africa and Latin America — are still having difficulty in launching their own programmes for social and economic development, often because of the effect of the superpower competition in their regions, it is remarkable that the ASEAN countries, already in the early 1970s, were conscious of the coming hazards and were trying to take concrete steps, however tentative, to maintain at least some measure of independence. Also, as Ghazali noted, they were pragmatic enough to realize that one of the prerequisites for political independence would be an ability to contain disputes within the region.

Admittedly, progress was slow. Even after they declared their intent in the Kuala Lumpur Declaration, the ASEAN nations were singularly inactive in defining what they meant by the 'initially necessary efforts', let alone translating them into negotiations with the great powers. However, in response to the accelerated pace of change in the region, ministerial conferences were convened far more frequently than before to discuss, often quite informally, a whole range of problems that member countries were facing. In this way, the five nations, despite their diverse backgrounds in terms of size, race, religion, history and political orientation, began to understand each other better, gradually generating a pattern of cooperation. A Thai intellectual has commented that this process helped to develop among the ASEAN countries a 'habit of consultation' and established the kind of relationship where one could 'agree to disagree, and to disagree without being disagreeable'.[35]

Chapter Four

JAPAN IN THE 1970s

The high hopes that Japan had had for the 1970s on account of its steady economic growth and increased international stature during the 1960s (expressed most clearly in the extremely successful Expo '70 held in Osaka) were rudely shattered. Its vulnerability was driven home by many events, notably the surprise announcement by President Nixon on 15 July 1971 of his intended visit to China, with its implications of a Sino-US rapprochement. Not only did this event — the first of the two 'Nixon shocks', as they were commonly known in Japan — constitute the most humiliating treatment that Japan had received from the United States since its defeat in the World War II, but it undermined the basic assumptions of its foreign policy during the 1960s. Also, it exposed the government's singular lack of insight into the changing realities of the international environment. In spite of its increasing contact with the world on account of its economic activities, Japan seems to have had little understanding of the circumstances which were forcing the United States to redefine the basic premise of its Asian policy, namely the containment of China. In a sense, it was a vindication of Japan's long-standing suspicion of the wisdom of that policy. Nevertheless, Japan was strangely reluctant to recognize the various signals presaging the change or to explore policy alternatives once it had happened.

Similarly, on the economic front, Japan was bent on maintaining the comfortable *status quo* which had facilitated its spectacular growth during the 1960s. The government continued to resist the liberalization of trade and investment and, in August 1971, attempted to ignore the effective collapse of the international monetary system. When, on 15 August, Nixon announced his new economic policy — the second 'Nixon shock' — Japan, instead of closing its exchange markets, as the European countries were doing, doggedly continued to buy dollars at their old par value of 360 yen — the Ministry of Finance assuming, in a gross miscalculation, that it could pre-empt the inflow of speculative dollars by its own exchange-control regulations. After holding out for twelve days, during which time $4.6 billion were purchased, Japan capitulated and on 27 August grudgingly introduced the flexible exchange-rate system.[1]

There followed a series of events that finally forced Japan to look more critically at its policies. The anti-Japanese movements which sprang up throughout Southeast Asia, beginning in Thailand in 1972, gave a jolt to a country that had little concept of the effect its economic activities was having on other nations. Then it was the turn of the oil shock to drive home to government and people the extreme vulnerability of their economy.

The purpose of this chapter is to follow some of these events which shook Japan during the 1970s, to speculate on why it was so unprepared for them, and to look at how, in spite of its often disorganized response, it did in fact manage to turn these circumstances into stepping-stones for subsequent growth.

Sino-Japanese Relations

President Nixon's announcement on 15 July 1971 of his intended visit to China provided all the countries in the region with a shock, but for the Japanese, who had not been consulted or forewarned, the effect of the US move was devastating. Only nine months before, President Nixon had solemnly agreed, in conversation with Prime Minister Sato, that 'both leaders . . . recognized the necessity for the two countries to coordinate fully their policy concerning China and agreed that they would maintain close communication and consultation about future developments on this issue.' And this promise had been reaffirmed in 1971 in a series of top-level US-Japan exchanges: for instance, between Secretary of State William Rogers and Foreign Minister Aichi Kiichi, Ambassador Armin Meyer and Prime Minister Sato, and Defence Secretary Melvin Laird and Defence Agency Director-General Nakasone Yasuhiro. These repeated reassurances blindfolded Japan to the move which was under way until literally 'a few minutes' before the President went on the air to make his historic announcement.[2]

Although Kissinger later admitted that it had been a serious error in manners and that there had been a lack of consideration to Japan, the Americans were hardly able to mitigate the psychological blow of having ignored Japan, which had thought of the United States as a partner and mentor, as well as a guarantor of its overall security.[3] Moreover, the blow was all the more severe in that policy towards China had been one of the most controversial issues of the 1960s, with Prime Minister Sato withstanding persistent public pressure for the normalization of relations with China and continuing to try not to disrupt the Americans' policy of 'containment'. Charges of misgovernment were levelled against the Japanese government, this time not only from the left but from a wide range of the population. People were now genuinely alarmed by the way in which the government had allowed the United States to treat Japan and, more importantly, by the fact that the United States and China could collude to shut Japan out of such an important development in the relationship between them. Prime Minister Sato's own political standing rapidly deteriorated as a result.

To some Japanese, the Sino-US rapprochement was an ominous reminder of the pre-war triangular relationship between Japan, China and the United States. The consistent and highly emotional support which the United States had extended to China in the early part of the century for the purpose of defending 'democratic' China against 'imperialist' Japan was a constant source of embarrassment to Japan and ultimately pushed it into the war in the Pacific. After the war, deeply disappointed by China's conversion to communism, the United States reversed its allegiance and began to support 'democratic' Japan and build it into a bulwark against 'communist' China. A surrealistic scenario seemed now to be unfolding, in which 'capitalist' United States and 'communist' China, newly allied, were squeezing Japan between their pincers. To make matters worse, the Chinese were at this time persistently alleging a revival of Japanese imperialism — a systematic propaganda campaign which helped to revive in the minds of the people of Asia memories of World War II and to arouse resistance against Japan's growing economic presence.

In spite of the deep humiliation and damage to his political career, Sato doggedly maintained his pro-US position for the rest of his tenure of office, supporting the abortive US initiative to defend the rights of Taiwan at the UN General Assembly in September 1971 and entering into negotiations with the United States on the textile dispute in October. He may have felt that it would be unwise to rock the boat at this point, when the return of Okinawa (which did, in fact, turn out to be another important development in the US-Japan relationship and a personal triumph for Sato himself) still lay ahead.

And, to rub salt into the wounds, China intensified its attacks on Sato's government, which it accused of militarism and economic imperialism. Zhou told a visiting Diet member, Kawasaki Hideji, that China would never accept a visit to Beijing by Sato, but that if a new prime minister were to replace him and make it clear that Beijing was the only legitimate government of China, he would be welcome.[4] This was an example of China's peculiar penchant for interfering in Japan's internal affairs by means of informal conversations with private individuals who visited China on invitation. Perhaps it was unavoidable as long as Japan persisted in not opening government-level relations in deference to US containment policy. The Chinese learnt to exert influence on Japan through various private channels, notably through the visits of academics, journalists, trade union leaders, women and youth groups who came to China with the government's tacit approval. Of course, the Japanese government used these same channels, when it wanted to communicate with China, although there was little Japan could do under the circumstances to influence China's policies.

As the conversation with Kawasaki shows, in spite of the shrill rhetoric against Japanese imperialism which China kept doling out publicly, it was becoming increasingly clear that in fact it wanted to normalize relations with Japan as soon as possible. Perhaps it calculated that it could use the US card more effectively against the Soviet Union when rapprochement with the United States was reinforced by normalization with Japan as well. Also, the process of negotiation with Japan might enhance its bargaining position vis-à-vis the United States. In any case, it was again people outside the government who were recruited to play the leading role in the process of developing the terms and conditions for the normalization. Kasuga Ikko of the Democratic Socialist Party (DSP), in April 1972; Furui Yoshimi of a pro-Beijing group of the LDP, in May 1972; Sasaki Kozo of the JSP, in July 1972; and Takeiri Yoshikatsu of the Komeito (Clean Government Party), in August 1972, were among those who visited China and undertook to mediate informally between the two governments. Through them, the government was able to put out a signal that it was not unwilling to accept the three principles laid down by Beijing as preconditions for talks on normalization: namely, (1) to recognize Beijing as the sole legal government of China, (2) to recognize Taiwan as part of China and (3) to abrogate the treaty with the Republic of China and sever diplomatic relations with that government.[5]

By mid-1972 the Chinese were obviously as eager as the Japanese for normalization. The visit of President Nixon to Beijing in February 1972 was not as rich in substance as the press coverage (which celebrated it as the 'event of the century') suggested, and, for all the talk of détente and a new balance of power, the United States proved a tough negotiator on the principles of the cold war, including the issue of Taiwan. The Chinese may have suspected that the ideologically less committed Japanese would prove more amenable to their wishes, if only because of historical and cultural affinities, reinforced by two decades of unofficial, yet active, people-to-people exchanges. In addition, the Chinese must have expected substantial Japanese economic help, which they needed rather urgently in order to proceed with their modernization programme. The Soviet attempt to lure Japan away from China and the United States by sending Foreign Minister Andrei Gromyko on a five-day visit to Tokyo in January 1972 must have given China an added incentive to hasten the process of rapprochement with Japan.[6]

In Japan, the enthusiasm for normalization with China, which had long been an unofficial national goal, was greatly reinforced now that it was regarded as the only way to restore the nation's self-respect and international standing, both of which had been seriously undermined by the policies of the United States. The sense of guilt, widely felt among the Japanese, about the damage which they had inflicted on the Chinese during the war must have further influenced the government to be as lenient as possible towards the Chinese position on normalization. In any case, liberated at last from the need to defer to the United States, the government itself was ready to move fast and far — even to outdistance the United States, it was hoped — in the subsequent relationship with China.

In May 1972 Okinawa was officially returned to Japan, thus setting the stage for the departure of Sato, who voluntarily resigned after eight years in office, by far the longest term in Japanese political history. This opened the way for the maverick Tanaka Kakuei to become prime minister on 5 July. Tanaka lost little time before plunging himself into a kind of normalization fever, thus strengthening the mood that already existed in the country. China, in turn, was quick to extend an invitation to the new prime minister, while giving an assurance, again through private channels, that even if they could not agree on the actual terms of a joint communiqué, this need not hinder the principle of establishing diplomatic relations.

Before going to Beijing, however, Tanaka visited Honolulu to meet with President Nixon on 31 August. A joint communiqué issued subsequently was significant in that it omitted to mention the Taiwan question, which was the main point of difference in the China policies of the two countries. The meeting seems, rather, to have confirmed the parting of their ways: namely, that Japan had already decided to sever official relations with Taiwan, whereas the United States was not yet ready to go that far. Thus Japan was finally able to reverse the position which had been forced upon it by Dulles in 1952.

On 29 September, a joint communiqué was signed by Prime Minister Tanaka and Premier Zhou in Beijing declaring that Japan had established diplomatic relations with the People's Republic of China. It also stated that Japan was keenly aware of its responsibility 'for causing enormous damages in the past to the Chinese people through war and deeply reproaches itself'. On Taiwan, Japan agreed with the position of the Chinese government that 'Taiwan is an inalienable part of the territory of the People's Republic of China'.[7]

Throughout the period leading up to Japan's normalization with China, the Foreign Ministry was bitterly criticized for having provided no information on the US overtures towards China, prepared no policy response to it, and taken little initiative in the subsequent process of normalization. Indeed it was difficult for the ministry to clear itself of the charge that it had allowed the United States to take Japan completely by surprise, and thereby to throw its decision-making system, at least temporarily, into total disarray. In retrospect, however, as long as its relationship with the United States remained an important anchor of Japan's position in the world, and in as far as the ministry was entrusted with the task of maintaining that relationship, it was in no position to initiate any move towards China — an action which would have been in breach of the specific agreement between Sato and Nixon to coordinate their policies. Also, in view of the dichotomy in Japanese public opinion, which has been sharply divided between pro-US and pro-China elements for many decades, the ministry had to be extremely circumspect about its actions, lest it precipitate further internal controversy.

The subsequent normalization with China had the effect of removing this political controversy and gave the ministry room for manoeuvre. The fact that China expressed its support for the US-Japan Security Treaty as well as for Japan's Self-Defence Forces had taken the wind out of the leftists' sails. Also, the normalization came at the end of a long line of problems with which the Foreign Ministry had had to deal arising out of Japan's defeat in the World War II, from the San Francisco Peace Treaty in 1951, the restoration of Japan-USSR relations in 1956, a series of agreements on the reparation of war damages at the end of the 1950s, the Japan-Korea Basic Treaty in 1965, through to the return of Okinawa in 1972. It gave the ministry a psychological boost that, from now on, it had a clean slate, as it were, and could consider its policies towards the world on their own merits without being constrained by the past.[8] Encouraged further by the removal of the ideological restraint of the US cold-war policy, the Foreign Ministry began to try out more assertive and imaginative moves in its own Asian policy, as seen in its overtures to North Vietnam during 1972, culminating in the establishment of diplomatic relations on 21 September 1973, and to Mongolia, which was recognized in February 1972.

Discord in US-Japanese Relations

Besides its impact on Japan's China relations, the 'Nixon shock' was symptomatic of a change which was to affect the basic framework of the US-Japanese relationship. That the United States was no longer willing or able to play the role of mentor/protector towards its Asian allies, including Japan, was made clear in the new orientation of US Asian policy as expressed in the Guam Doctrine. This obviously meant that, from now on, Japan would be expected to play a more positive and assertive role, if only because of its growing economic power. Unfortunately, Japan seems to have been unable to orient itself in the emerging new environment of the region, or to understand the role that was expected of it. In his foreign policy report — or 'State of the World' address — to Congress of February 1972, President Nixon could not disguise his impatience with Japan's apparent passivity, when he commented on the 'shock' which he had caused to the country in the previous year:

> We recognize that some of our actions during the past year placed the Japanese government in a difficult position. We recognize that our actions have accelerated the Japanese trend toward more

autonomous policies. We regret the former but could not do otherwise. We welcome the latter as both inevitable and desirable — inevitable because it reflects the reality of Japanese strength in the 1970s — desirable because it is a necessary step in the transformation of our relationship to the more mature and reciprocal partnership required in the 1970s.[9]

Difficulties in the US-Japan relationship had started as far back as January 1968, when Japan refused an American request that it purchase mid-term US treasury bonds as part of a scheme to defend the weakened dollar. At that time, largely owing to the prolonged war in Vietnam, the American economy was suffering from the persistent trilemma of slow growth, inflation and growing unemployment. Despite the fact that Japan's GNP had almost trebled in the decade of the 1960s, it refused the US request on the grounds that its foreign exchange reserve was still hovering at a meagre $2 billion level. Economically the decision may have been a sound one. Also, given that public opinion was still largely dominated by anti-Vietnam sentiment, it would have been difficult for the government to explain any move which in effect would have been helping the United States with the cost of its war. However, in refusing, Japan lost a precious opportunity to get on to a partnership footing with the United States, the lack of which was to cause it considerable problems later on.

Japan's difficulty in relating to the rest of the world appears to have applied particularly to political issues. This was perhaps because of the pervasive influence of economism during the 1960s. When they met with political pressure from outside, the Japanese seem to have had difficulty in understanding what lay behind the pressure and, as a result, in formulating a coherent and consistent response to it. The so-called 'textile wrangle' of 1969–71 is a case in point. This began in May 1969, when the visiting US Secretary of Commerce, Maurice Stans, requested the Japanese government to take the necessary steps to restrict the export of textiles to the United States in order to fulfil the election platform which President Nixon had pledged in his 1968 campaign. For Japan, this was tantamount to being asked to pick up Nixon's personal political bill. Failing to grasp the political implications of so extraordinary a request, the Japanese government responded to it on purely economic grounds: in short, it conveyed its refusal to Stans, giving as its reasons that there was little evidence of actual damage being done to the US textile industry, and that to agree to such a demand would be to violate the principle of free trade to which both countries subscribed. In view of mounting opposition throughout the country, notably among the small-scale textile manufacturers, it was perhaps the only choice for the government. Nevertheless, by refusing the American request and citing, rather self-righteously, the principle of free trade, Japan missed the chance to adapt its behaviour to the USA's new attitude towards it, which was to be less protective and more demanding of Japan.

The new framework of US policy in Asia as propounded by the Nixon administration, did not, however, always present problems for the Japanese. For instance, to their pleasant surprise, it worked to their benefit on the question of Okinawa, particularly in respect of the nuclear-free status which they wished to maintain for it after its return. Contrary to their apprehension, when they broached the subject with the United States in September 1969, the United States readily concurred and, on Sato's visit to Washington in November 1969, it was agreed that Okinawa would be returned to Japan in 1972, nuclear-free. The Nixon administration may have wanted to demonstrate by such generosity that it meant what it had said in the Guam Doctrine, that from now on the United States would provide its Asian allies with only an ultimate nuclear umbrella, while expecting the Asians themselves to be responsible for conventional warfare. Also, in the context of encouraging self-reliance in the Asian countries, it may have calculated that it would be counterproductive to become embroiled again in Japanese domestic controversies by rebuffing Sato's request at this juncture. In any case, it was a bold move on the part of the United States, unthinkable only a few years previously, which may well suggest a personal intervention by President Nixon himself.[10]

There was a persistent rumour in Japan that when Nixon and Sato met in Washington, they had come to a secret understanding that, in return for the US concession on Okinawa, Japan would acquiesce in US demands on the textile issue. This lent a sensational character to the problem and hardened the determination of the textile industry not to give in to government pressure. It is difficult, however, to imagine that American policy-makers had conceived the concession on Okinawa as being directly linked to the textile dispute. The background and context of the two issues were too different to be equated on the same plane. However, having once decided upon the concession on Okinawa, it is quite likely that the United States expected Japan to reciprocate in some way or other. Although the term 'free ride' was not yet in circulation, the United States had long felt that its relationship with Japan had been too one-sided, and that it was about time that Japan be made to pay a price, however modest, for its alliance with the United States.

By that time, however, it was difficult for the Japanese government to change its economistic, as well as its self-righteous, posture and persuade the industry to give in on purely political grounds. As a result, in spite of numerous talks and negotiations of every conceivable kind — government-to-government, Diet-to-Congress, industry-to-industry — the problem remained unresolved, accumulating still further ill feeling on both sides. There was little that the Japanese could do to dispel long-standing American resentment at the way in which Japan used its dubious economism as a smokescreen, refusing to give even minimum cooperation to the United States and amassing its wealth while the United States was bleeding in Vietnam. One indication of the strength of US resentment is a report that the administration had played with the idea of applying the 10 per cent surcharge on US imports (which was to be part of Nixon's new economic policy) exclusively on imports from Japan.[11]

It is a pity that such an intractable problem as the 'textile wrangle' should have provided the Japanese with their first taste of the new orientation of US Asian policy. The textile dispute was too small an issue in itself and yet so emotion-ridden that the Japanese tended to see only the ruthless side of the new US administration and to miss the far-reaching implications of its policy for the Asian Pacific region. The 'Nixon shock', too, was regarded primarily as an expression of American ill will towards Japan, rather than as a symptom of a new pattern of relations in the region. It was unfortunate that, until — in October 1971 — Japan finally capitulated and accepted the US demands on the textile issue on political grounds, the 'wrangle' dominated US-Japanese relations to such an extent that there was little opportunity for the two governments to conduct a dialogue on more important issues, such as possible areas for future cooperation.[12] In fact, however, since Japan was not yet prepared to think in global terms, particularly in a political context, it would perhaps not have been able to undertake a sensible dialogue with the United States on global strategy, while the United States, for its part, was too unilateral in its attitude to make a dialogue really fruitful.

* * *

The 'textile wrangle' was merely the first in a series of the US-Japan trade disputes which lasted throughout the 1970s. It is perhaps worth looking briefly at three particular product issues: colour television sets, steel and cars. In the mid-1970s, the American television industry appealed to the US International Trade Commission (ITC), attributing its job losses to a rapid increase in colour television imports from Japan (in 1976 growing by 150 per cent over the previous year and taking 40 per cent of the US market). Unwilling to repeat the textile experience, the Japanese responded rapidly by negotiating, in 1977, an orderly marketing agreement (OMA), which was to last for three years. To offset these export losses, Japanese producers greatly increased their production facilities inside the United States, and thus succeeded in keeping their exports well within the shipment quotas set by the OMA. The shortfall has, in fact, been taken up by other Asian television producers, notably South Korea and Taiwan, with which the United States subsequently moved to conclude OMAs.

As regards steel, as the US market recovered in 1976-7 from the post-oil-shock recession, Japanese steel exporters were the major beneficiaries. Pressure from the US steel industry and Congress to curtail Japanese imports and bring anti-dumping suits led the Carter administration to work out a trigger price mechanism (TPM), based on Japanese production costs, in return for the withdrawal of the suits. Since 1978 the Japanese share of the US steel market has declined slightly (it is now the Europeans who are in conflict with American producers) and is likely to continue to do so as the Japanese diversify their export markets.

Finally, there is the case of the small and energy-efficient Japanese cars that made such significant inroads into the American market following the second oil shock (in 1979 three out of every four imported cars were Japanese). The issue became politicized in 1980, when the United Auto Workers, the largest union involved, and the Ford Motor Company protested. Although the ITC rejected claims that Japanese imports had 'seriously damaged' the US car industry, the Japanese had to agree in 1981 to restrict their car exports to the United States by 7 per cent.

These disputes, though economic in origin, inevitably strained political relations, since they provoked increasingly strong pressure tactics from the Americans. On one hand, the United States, using every means at its disposal (OMAs, TPMs or outright export restriction), was urging export restraint upon Japan, while, on the other, it was calling for the liberalization of its market and pressing for the removal of import barriers. It was particularly critical of the so-called 'non-tariff barriers', arguing that Japan's distribution system, product standards, government procurement policies and technical regulations were unnecessarily complicated. Such criticism was often extended to Japan's macroeconomic policies, and to its social and cultural structure.[13] Towards the end of the decade, American politicians began to suggest a linkage between economic and defence issues, and to call for a larger defence expenditure, commensurate with Japan's economic power. Given its overall dependence on the United States, these new elements in the relationship between the two countries were extremely difficult developments for Japan to handle.

Anti-Japanese Movements in Southeast Asia

On 20 November 1972, a ten-day drive to boycott Japanese goods was launched in Bangkok at the initiative of the National Student Centre of Thailand (NSCT), an organization based in Chulalongkorn University, claiming a nationwide membership of over 100,000 students. The purpose of the boycott, according to the NSCT, was 'to make the Thais realize the danger of the Japanese economic domination of their country'. The students distributed posters and stickers to various parts of the country, urging the populace not to buy Japanese goods. It was the beginning of a series of anti-Japanese demonstrations which were to spread through Southeast Asia for the next year or so, leaving an indelible mark on its political landscape.

Thirayut Boonmee, then chairman of the NSCT, cited as one of the grievances of the students, 'the huge trade deficit that Thailand had with Japan and the feeling that our business sector is dominated by Japanese businessmen'. This reaction against Japan's domination of the Southeast Asian economy was strong also among the older generation of Thai intellectuals.[14] Except for a few years of occupation during World War II, Japan's presence in Thailand and in Southeast Asia had never been more pervasive, and what was most disturbing was that there seemed to be no way to reduce it. Besides being the biggest market for its primary commodities as well as the biggest supplier of capital goods, Japan was flooding the daily life of Southeast Asia with every conceivable item of consumer goods. One student wryly commented that even the anti-Japanese posters which they were busy distributing were made of Japanese paper, and were printed by Japanese printing presses, using Japanese ink.

Also, bitter criticism was levelled by the students against the 'colonial life-style' of Japanese businessmen, who were aloof and arrogant, living in their exclusive enclaves, adopting discriminatory attitudes towards local managers and workers. Similarly, the apparent unwillingness of Japanese executives who were operating in joint ventures to promote local partners to managerial positions was cited as evidence of Japan's chauvinism as well as of its reluctance to transfer technology to its Thai partners. The Japanese were portrayed in general as being callous and heartless, interested only in exploiting the resources and markets of Southeast Asia without giving back adequate benefits to the region or its people in terms of social and economic development.

Seeing the surge of anti-Japanese sentiment around the country, the Thanom-Prapas government, which had faced a variety of difficulties since the Sino-US rapprochement, and particularly since the bloodless coup in November 1971, chose to lend its support to the students. In fact, this move proved to be a double-edged sword, because the students' criticism of Japan was subsequently switched to criticism of the government, which, after all, had allowed the Japanese to behave as they did. Perhaps, however, the government calculated that, by appearing to support the students, it could obtain concessions from Japan and use these to boost its flagging image in the country. Commenting on the students' rally on 20 November, General Prapas Charusatien, the deputy chairman of the ruling National Executive Council, supported the students by saying that 'they were acting within their rights when they wanted the Thais to stop buying Japanese goods to remedy the situation'.[15] Encouraged by the government's attitude, the movement soon spread to many parts of the country.

This sudden outbreak of fiercely anti-Japanese sentiment caught Japan completely by surprise. Obviously it was a serious drawback to find so much hostility in Southeast Asia just at the time when the relationship with the United States was showing increasing signs of strain. Also, unlike the difficulties with the United States, the events in Thailand were so unexpected that the Japanese hardly knew what had hit them. At a loss what to do, the government hastened to send Nakasone Yasuhiro, then Minister of International Trade and

Industry, to Thailand in January 1973, partly to find out what was going on and partly to mollify the feelings of the Thais by showing a sincere intention to rectify whatever it was that was antagonizing the Thai public. As an opening move, Nakasone offered a package of concessions, including the untying of loans hitherto tied to the purchase of Japanese goods, the intensification of efforts to remedy the trade deficit and further assistance for Thailand's economic development. Later on, he even went so far as to apologize to the Thai government, saying that 'the Japanese government regrets some actions of Japanese people here. We will try to correct the situation.'[16]

As the rhetoric of the students and the unmistakable hostility of the public were reported back to Japan in detail, the Japanese realized that they had been too complacent about the political implications of their growing economic activities in Southeast Asia. After World War II, they had pledged themselves never to become involved in political, let alone military, activities abroad, and to confine themselves to economic matters. It was generally and naively believed that, as long as they steered clear of political and ideological entanglements, the economic involvement alone would not arouse any political controversy. The events in Thailand made Japan realize, for the first time, the enormous size and scope of its economic presence in Southeast Asia as well as the political responsibility of carrying on such extensive activities in other countries. A Thai scholar likened Japan's impact on the economy of Southeast Asia to the entry of a big warship into a quiet little port. Whenever the warship changed course, or even made the slightest movement, the small boats and sampans that filled the port were seriously affected, some even capsizing, with little or no awareness on the part of the warship that had made the move in the first place.[17]

The sudden surge of anti-Japanese movements in Southeast Asia had to do with more than Japan; it was also related to the change in the political environment in general. When the war in Vietnam was in full swing, Japan's activities did not attract much attention. However, with the prospect of American withdrawal so near, Japan's visibility greatly increased. Furthermore, the newly articulated concept of a multipolar world, with Western Europe, China and Japan to be added as extra power-centres to the previous bipolar system, naturally stimulated a fresh interest and concern about Japan and its political intentions. Finally, the warnings which China had been persistently broadcasting during 1970–2 about the revival of Japanese militarism/imperialism certainly helped to reactivate old fears of Japan in Southeast Asia.[18]

In the meantime, the NSCT, which had picked up considerable momentum through a further series of anti-Japanese demonstrations, moved towards a showdown with the powerful Thai military government in October 1973, with protests against the arrest of students escalating into violent clashes between police and students. This led the King to withhold his support from the Thanom-Prapas government, which fell from power as a result. The students' anti-Japanese fervour, however, was by no means abated. In early January 1974, when Prime Minister Tanaka arrived in Bangkok for a state visit, he was met at the Dong Muang airport by students carrying placards which bore such slogans as 'Get out you ugly imperialist' and 'Imperialist monster Tanaka'.[19] The government of Prime Minister Sanya Dhamasak tried to keep the situation under control, but, being a caretaker administration, it did not have the authority to curb the students, fresh from their triumphs over the previous government.

But this was not all. When Prime Minister Tanaka and his entourage touched down at Jakarta on 14 January 1974, the whole city was ablaze with student rioters who, encouraged by the latest achievements of their Thai counterparts, rampaged the streets. 'Troops,' according to a press report, 'had to force a path through chanting student demonstrators . . . to make way for a motorcade carrying visiting Japanese Prime Minister Tanaka into the city.'[20] The following day the situation turned worse and began to look like a full-scale revolution: 'Cars were set ablaze, the Rising Sun was lowered from the flag masts of government buildings and troops fired over the heads of the mob.'[21]

The riot, which was later christened Malari (a contraction of Malaputaka Januari, or 'January Misfortune') exposed a deep division in the leadership of Indonesia and turned out to be the most serious challenge to the Suharto government since it came to power. In contrast with the situation in Thailand, the real target of the riots in Indonesia was not so much Japan as the government itself. The students, along with intellectuals and other like-minded people, had long been impatient that the fruits of economic development, the centre-piece of Suharto policies, rarely found their way to ordinary citizens. A professor from a leading university in Indonesia who was implicated in this riot and spent a few years in prison expressed surprise when he was asked on release whether his feelings against

Japan were still so strong; at no time, he said, all through this struggle, had Japan ever been an issue in his mind.[22] Similarly, the *New York Times* reported that 'the Japanese were but the most visible and accessible target at the moment — more accessible than the high Indonesian officials who were widely suspected of corruption but whose power protected them against direct attack.'[23]

Nevertheless, the sheer force of the violence and the ugly mood of the rioters in Jakarta shook the Japanese to the core. Even if the riots were really targeted at the Indonesian government, it was clear that Japan was fully implicated in whatever the demonstrators were opposing. For example, there was a lingering rumour of corruption in high places as well as alleged collusion between Indonesian power elites and Japanese business interests. On the pretext of not interfering in the politics of other countries, Japan chose not to pay much attention to such allegations. However, these were the cases which represented the 'political responsibility' that Japan was expected to fulfil if it was to sustain its economic involvement in Southeast Asia.

It is frightening that Japan utterly failed to predict the outbreak of such incidents in Indonesia until the prime minister's party was engulfed at the airport in Jakarta. This showed a singular insensitivity on Japan's part to the changing political realities in Southeast Asia, notably the role of intellectual elites in determining the mood of the nation. Thanks to generous philanthropic as well as governmental efforts on the part of the United States and West European countries, a substantial number of intellectuals who had received higher education abroad were beginning to take over the universities and various government agencies in both Thailand and Indonesia. These people were in the vanguard of the movement to modernize the political system and as such provided the conceptual framework for the political activities of the students.

The one saving grace in all this was the fact that the blatant and often quite rude display of anti-Japanese emotions in Southeast Asia did not stir up narrow nationalistic reactions in Japan. There were surprisingly few expressions of anger, even when the prime minister was exposed to outright humiliation in foreign lands. Also, it was fortunate that Japan did not put too great a political interpretation on these events. An economistic Japan had certain drawbacks, but a Japan that was hypersensitive politically could well have been more unsettling to the region. In any case, the government launched a programme to promote better understanding of Southeast Asia among Japanese intellectuals and businessmen, and to work out a desirable mode of behaviour. Similarly, in Thailand and Indonesia, the fever of anti-Japanese sentiment soon died down under the weight of more pressing political problems; in fact, it hardly left any serious scars at all on the region's subsequent relations with Japan.

The Oil Shock

On 26 September 1973 Prime Minister Tanaka left Tokyo to travel to France, Britain, West Germany and the Soviet Union. The purpose of this tour by Tanaka, the man who had put forward the ambitious idea of 'remodelling the Japanese archipelago' and who held an extremely positive economic vision for Japan, was to promote diversified resource diplomacy. Already at this period Japanese business circles were anxious to diversify the sources needed to meet the country's increasing demand for raw materials and energy. Tanaka's visit, therefore, was intended to support these diversification efforts by talking with the French about the purchase of uranium concentrate, and with the British about Japanese capital participation in the development of North Sea oil, before going on to Moscow via West Germany. The most important item on his agenda with the Russians was to make a start on negotiations for the return of the Northern Territories, and Tanaka had conceived of possible Japanese participation and cooperation in the Tyumen oilfield development as a bargaining counter. Ironically, however, on 6 October, the day before he reached Moscow, the Arab-Israeli war broke out, an event that altered the economic structure of the whole world, to say nothing of putting a brake on Tanaka's resource diplomacy.

At first, most Japanese, in common with most of the rest of the world, thought that Israel would again display its overwhelming strength and the war would soon be ended. However, there were officials at both the Ministry of International Trade and Industry (MITI) and the Foreign Ministry who were aware that this war might affect the supply of oil from the Middle East. From about 1970 onwards, the traditional domination of the world oil business by European and American business interests had begun to be undermined by growing oil nationalism (demands for price rises as well as for the nationalization of foreign oil companies), led at first by Libya, then followed by Saudi Arabia. Moreover, in September 1970 King Faisal had declared that Saudi oil production would be reduced by one million barrels a day until the United States changed its Middle East policy. This was a clear indication that oil was going to be used as a political weapon, but the Japanese had little idea of what provision to make for such a situation. Over 60 per cent of Japan's oil imports were supplied by the European and American oil majors, and as a result the Japanese knew next to nothing about the actual processes of extraction, refining and distribution. Japan, therefore, was hardly ready for the effects of the Arab-Israeli war of 1973.

Contrary to general expectation, the Israeli army and air force suffered great losses in the Sinai peninsula at the hands of the Egyptian army, which was equipped with Soviet missiles; in the following week Iraq, Sudan, Jordan and Saudi Arabia joined in the war against Israel on the Egyptian-Syrian side. Then, on 17 October, at a meeting of the oil ministers of the ten-nation Organization of Arab Petroleum Exporting Countries (OAPEC), it was decided to reduce daily oil production by 5 per cent until Israel withdrew completely from those areas occupied in 1967; to place an oil embargo on the United States and the Netherlands; and to guarantee previous levels of supply only to 'friendly countries'. Japan, amid deepening uncertainty and impotence, realized that — unlike Britain, France, Austria and many African countries — it did not fall into the category of 'friendly countries': those countries that gave military aid or exported weapons to the Arab countries, or had broken or had no diplomatic relations with Israel. On 20 October the major oil companies announced a substantial rise in the price of their oil, together with a 10–30 per cent reduction in supply. At one stroke, the basic assumptions underlying Japan's economic goals for the 1970s were destroyed.

Japan was particularly shocked that, despite buying huge amounts of Arab oil, it should have been classified, without a single word from the Arab side, as an 'unfriendly' country. Perhaps it was seen as pro-American and pro-Israeli, and even though it was buying a markedly greater volume of Arab oil than Britain or France, it was clearly not regarded as having any importance, either as a market or as a political partner. This was a crucial verdict on the inadequacy and clumsiness of Japan's resource diplomacy and Middle East policy; perhaps a warning, too, for the economistic emphasis that tended to dominate all its relationships.

On 9 October, President Nixon, anticipating the likelihood of an oil crisis, had appealed to the nation to conserve oil, and three days later announced the introduction of rationing. Other countries, such as Britain (which had already experienced rationing during the war and at the time of the 1956 Suez crisis), were proceeding with rationing and conservation measures. But in Japan's case, where 99 per cent of all oil was imported, MITI had little control over the system of supply and distribution, and it was highly doubtful whether a rationing system could actually be implemented even if it became necessary. Japan's sense of unease and impotence was therefore heightened by the feeling (though this may not have been based on fact) that it was lagging behind the Western countries in this particular respect. The public seemed to sense that the government was being laggardly, and during November wild rumours of shortages of oil-related products swept the country. People hunted the shops for such goods as kitchen detergent or toilet paper, and there was even a run on one of the local banks.[24] As Ronald Dore has remarked, it seems that in a consensual society such as that of Japan, 'shared anxieties can rapidly develop into a sense of crisis by the reverberatory process of cumulative feedback' — a process that culminates in a state close to a 'national nervous breakdown'.[25]

As a preliminary measure, the Japanese government tried to safeguard the volume of oil imports by buying on the spot market and through direct deals with oil-producing countries, while — with the long term in mind — it concentrated on working out measures to enable Japan to join the OAPEC category of 'friendly countries'. Since the 1960s the Foreign Ministry had been training a nucleus of Arab specialists and trying to gather information on the region, but, since the oil deals were virtually monopolized by major oil companies, there was little direct involvement.

In the area of political involvement, Britain had a strong foothold in the region in the sense of knowledge and contacts acquired in the colonial period. In the case of the United States, although temporarily one of the targets of the OAPEC embargo, there was no denying its deep relationship with Saudi Arabia through military aid, infrastructure assistance, large purchases by the Aramco oil company, and substantial Saudi capital investment in the United States. Moreover, in view of its position as a superpower as well as its influence over Israel, the Arab countries certainly could not simply ignore it. By comparison, Japan had no bargaining position at all vis-à-vis the Arab countries. There had been little cultural interchange and the Arab side had very little knowledge of Japan. In view of the fact that Japan depended on these countries for over 50 per cent of its oil — the very life-blood of its economy — this was a frightening situation.

The Japanese government began to re-examine its past policies towards the Middle East and to seek the limits to a pro-Arab policy within which Japan could preserve its dignity as a country and maintain balance in its relations with other countries. The Foreign Ministry enlisted the help of former diplomats and businessmen who had worked in the Middle East and gave them the task of gathering as much information as they could about what measures were necessary for Japan to obtain 'friendly country' status.[26] However, since relations with the Middle East were closely connected with relations with the United States, Japan was debarred from moving unilaterally towards a blatantly pro-Arab position. In fact, en route back to Washington from a visit to Beijing, Kissinger, Secretary of State, called in at Tokyo and talked with Tanaka on 15 November specifically on the question of the Middle East. He said that the United States was working on a peace plan for the Middle East and asked Japan to avoid any independent action. Reportedly, by way of return, Tanaka asked whether the United States would guarantee Japan's oil supplies if it followed American wishes.[27]

At an OAPEC oil ministers' meeting in Vienna on 18 November, all the EEC countries, with the exception of the Netherlands, were exempted from the 5 per cent supply cut, but again not Japan. Perhaps, the Arab side wanted to restore friendship with the European countries with a view to obtaining their future political support. If so, it was another blow to Japan's international standing not to be considered.

As if prompted by this, on 22 November the Chief Cabinet Secretary issued a statement reiterating Japan's policy towards the Middle East. In it, the government reconfirmed its adherence to Resolution 242 of the UN General Assembly (in which Japan had expressed understanding for the Arab position); suggested guidelines for a Middle East policy which included Israeli withdrawal from all territories occupied since 1967; and deplored the Israeli occupation of Arab lands, adding that Japan would have to reconsider its position vis-à-vis Israel pending further developments in the region. These three points were based on such information as the government had collected so far, and represented the limit to which Japan could go towards a pro-Arab position in the circumstances.

Against the background of this new diplomatic stance, the government sent the deputy prime minister, Miki Takeo, as special envoy to eight Middle East countries (Saudi Arabia, Egypt, United Arab Emirates, Kuwait, Qatar, Syria, Iran and Iraq) in mid-December. His reception in Saudi Arabia was quite warm, and on 12 December King Faisal promised to give his full cooperation in securing Japan's designation as a 'friendly country'. However, the Saudi government took this opportunity to request Japanese assistance on a variety of projects, including an oil refinery, a petrochemical plant, a steel-mill and railway construction.[28] In Egypt, Miki promised to give a loan of $280 million for the reconstruction of the Suez Canal.[29]

The countries in the Middle East were neither particularly hostile nor particularly friendly towards Japan. The problem was that there had been insufficient diplomatic effort in the past, mainly because the relationship had been effectively dominated by oil, which was traded largely through the Western oil interests. It was clear that Japan would have to begin building up new bilateral relationships with these countries, and would perhaps have to pay a price for it. Nevertheless, the government's statement and Miki's tour did serve the current purpose, for on 25 December, as Miki was leaving Iran for Iraq, a meeting of the OAPEC oil ministers in Kuwait recognized Japan as a 'friendly country' and agreed to restore the volume of oil to Japan to September levels.[30] However, it was revealed at the same time that, two days before, a ministerial meeting of the six Gulf countries, also held in Kuwait, had decided to raise the price of crude oil by 228 per cent, from $5.11 to $11.65 a barrel.

Price, rather than the maintenance of supply, now became the prime concern. The 'panic' that had lasted for the past several weeks changed from a fear of shortage to a fear of inflation, the rate of which, even without the rise in the price of oil, had been steadily mounting since the end of 1972.[31] The immediate problem was in what way the Japanese economy, used to rapid growth, could decelerate, and how it could adapt to a price structure based on the new price of crude. Unfortunately, the process was complicated by the steadily declining popularity of the Tanaka cabinet, a trend which was accelerated by the oil shock. Obviously, Tanaka's expansionist economic policy had lost its rationale as a result of the oil shock, and the falling level of support for him is reflected in opinion polls conducted at that time — dropping from 62 per cent in August 1972 to 22 per cent in November 1973.[32]

In order to put a brake on its declining popularity, the government took measures to lower the price of oil for domestic consumption, an expedient which merely led to a rise in the price of oil for heavy industrial goods. In 1974 consumer prices in Japan rose by 24–25 per cent, a rate considerably higher than that in the Western industrialized countries. While this might have been considered natural given Japan's high dependence on imported oil, at the time it was felt that, by comparison with the other industrialized countries, Japan's response to the oil shock had been inept. The public's confidence was undermined, and its sense of vulnerability reinforced. As a result, Tanaka's popularity never recovered, and in December 1974, in connection with an allegedly shady financial deal in his political organization, he was forced to resign, to be succeeded by Miki.

With the benefit of hindsight, one can argue that Japan's response was not necessarily so inept. In fact, Japanese industrial cost-efficiency improved, and its export competitiveness subsequently rose so dramatically that the oil shock could even be considered a blessing in disguise. This can be attributed not only to the macroeconomic policies which the government set in motion, but also to the efforts on the part of Japanese industry to save energy and costs. Japan was fortunate by comparison with other industrialized countries in that industrial energy consumption was very high and domestic consumption very low (see Table 4.1). This demonstrated that, for all the high growth in per capita income in the 1960s, the habits of a thrifty life-style from earlier, poorer times remained strong. As a result, when the government took measures to encourage energy-saving, a significant improvement was achieved.[33]

Table 4.1: 1970 Oil Usage (%)

	Japan	USA	W. Germany
Domestic	21	34	33
Industrial	59	31	42
Transport	13	22	12
Energy Production	7	13	13

Source: Yanagida Kunio, *Okami ga Yattekita Hi* (Tokyo, 1983), pp. 283–4.

In Japan both the government and the people have had plenty of experience in responding to emergencies, such as the coal industry's crash programme of rehabilitation after the war or the development of certain key export industries. To quote Ronald Dore again, this ability to respond to a challenge is central to the Japanese national character:

The sense that the Japanese nation faces an uncertain and on the whole hostile world, that all Japanese must cooperate to meet some impending challenge to the nation's integrity or honour or prosperity, is one that has been continously with the Japanese ever since the 1870s when they embarked on their career as late-developing, catching-up Asians in a Euro-American world — and has, indeed, been one of the important ingredients of the success of that career.[34]

When, in November 1973, the government reviewed the situation, it decided that the answer to Japan's oil crisis was not rationing, but energy conservation. MITI therefore set in motion a drive for energy conservation throughout industry, while at the same time allowing market forces full freedom. A comprehensive programme was developed for educating the public; promoting research and development, both on energy conservation technology (the 'moonlight' plan) and on alternative energy (the 'sunshine' plan); and giving financial inducements for energy-saving procedures and products. Business circles, aware that conservation and cost reduction were the only way out if the economy was to survive in the face of the paralysing effects of a threefold rise in the price of oil (in 1974 Japan experienced negative GNP growth for the first time since the war), cooperated fully in this programme. Japan's particularly homogeneous and goal-oriented social structure favoured the implementation of any effort of this kind. The government knew that it could rely on the firms' cooperation in the new energy programme; the firms, in turn, could count on all their workers, both white- and blue-collar, to respond to the company's requirements in the consciousness that they were in the national interest.[35]

The volume of Japanese imports of crude oil decreased in 1974 and again in 1975, and in 1976 increased by only 1.7 per cent, even though real GNP grew by 6 per cent. Japan also looked to diversify its sources of oil and other energy supplies away from the Middle East; this inevitably increased interest in sources closer at hand, in the Asian Pacific region. Japan looked increasingly towards China, resulting in agreements in 1977 and 1978 for China to export 47 million tonnes of crude oil to Japan over a five-year period. The Japanese also became involved in obtaining supplies of LNG (liquefied natural gas) from Indonesia, Brunei and, later, Malaysia.[36]

As a result of these various efforts, Japan was able to accomplish much more than its original energy-saving drive had set out to do, and even effected a thorough-going revolution of Japanese technology. The sense of urgency and threat induced by the oil crisis forced the Japanese to re-examine all aspects of the economy, which had fallen into a kind of hypertrophy and complacency after the growth of the 1960s, and to take a new look at the rationalization and cost-effectiveness of processes at both company and factory level. As part of the energy-saving programme, numerous technological innovations had been developed, and these, backed up by input from new computer and electronic systems, led to a comprehensive rationalization of Japan's production technology. The consequent rapid progress in export competitiveness became one cause of the trade friction that has continued into the 1980s. Also, the qualitative improvement in Japan's economic activity enabled it to respond effectively when the second oil shock came.

The oil shock of 1973 throws interesting light on Japan's unique position at that time. Despite its widespread economic activities, it was still virtually cut off from the mainstream of international knowledge and experience. The European countries had interests in common and experience of common action, so that they could decide on a position while considering each other's responses. Their relations with the Middle East were difficult because of the element of anti-Europeanism involved, but, to make up for it, they had a stockpile of collective knowledge and experience of the region dating from colonial times. In Japan's case, everything had to be carried out alone. This merely reinforced its sense of vulnerability and isolation. Even the technological improvements carried out under the pressure of the oil shock would not have been possible if it had not been for the people's acute sense of crisis, a 'save energy or die' mentality. However, they were rewarded with fruits greater than they had bargained for: they were aiming at cost reduction and achieved overcompetitiveness. The advances in production technology which Japan began to demonstrate after the oil shock frightened the United States and provoked a negative reaction in Europe. In this sense, too, the oil shock of 1973 was a turning-point for Japan.

Chapter Five

A NEW PARADIGM OF RELATIONS

There is little doubt that the rapprochement between the United States and China altered the relative positions of the superpowers in the region, with the Soviet Union turning out to be the net loser. The end of China's isolation from the world, as indicated by its entry into the United Nations, and the normalization of relations with many countries including Japan, greatly complicated the Soviet Union's planning in the region and reduced the influence which it might have wielded there. If China conceived the whole move as a countermeasure to the Soviet Union's hegemonic stance as symbolized by the Brezhnev Doctrine, then it was a brilliant success.

The winner seems to have been the United States, which found itself in a much stronger position than before. In a triangular relationship, the one who is on talking terms with both of his adversaries is likely to gain the most advantage. Given the Soviet Union's fear of the US-China relationship growing any closer, the United States now had a pull on both China and the Soviet Union by virtue of its role as balancer between them. It was perhaps the first step towards Kissinger's version of the balance of power: a system whereby a group of major international actors are engineered into working together so as to maintain a framework of orderly management in the world.

In the Southeast Asian context, this meant that the United States could withdraw some troops from Vietnam without losing too much prestige. Indeed, although the war was still being fought in Vietnam, often quite fiercely, and the United States was continuing to bomb the North, the conflict was quickly losing the appearance of a frontline contest between the superpowers. Even the role of the United States had changed: from being one of the main contestants, it had become a self-appointed referee, willing to mediate, by force if necessary, the conflict between North and South. In view of all it had gone through because of the war in Vietnam, and the length of its involvement, this new role went some way towards mitigating the damage of its past failures.

Despite the Sino-US rapprochement and the signing of a peace agreement in Paris in January 1973, allowing for the withdrawal of US forces and a cease-fire in Vietnam, the actual fighting did not end. Moreover, the United States was powerless to prevent the large-scale attacks, made by the North Vietnamese and the revolutionary forces inside South Vietnam in March and April 1975, that resulted in the destruction of the South Vietnamese government. On the other hand, one can perhaps argue that if it had not been for the Watergate crisis which caused acute paralysis at the very heart of the US decision-making machinery, such a speedy and humiliating capitulation might have been avoided.

In any event, what the collapse did demonstrate was the fragility of the national base of South Vietnam, which, in the last few months under the Thieu government, had suffered confusion, corruption and a process of rout, and therefore how difficult it had been for the United States to maintain its efforts over the preceding years. Ironically, however, the situation that finally emerged, — namely an environment that was conducive to regional development — was quite close to the USA's original objective, even though it had been reached via a different route.

Although the fall of Saigon, on 29 April 1975, and the final withdrawal of the Americans, had been expected for some time, the reality of the communist victory in Indochina did have a great impact on the region. It provided the decisive stimulus for the formation of a new international order in Asia. Organizations such as ASPAC and SEATO, whose objective had been to build cold-war-oriented and great-power-connected systems in the region, had lost their momentum. (Pruned down in 1973, SEATO was finally dissolved in 1977, although technically the original treaty remains in force, while ASPAC had no more meetings at Council level after 1972.) Although it was thought that a multipolar system — in which the United States, the Soviet Union and China, as well as the economically powerful Western Europe and Japan, would play prominent roles — would modify the now increasingly shaky bipolar system, nobody knew how such a system would function. Ironically, in contrast with this increased multilateralism at a global level, Southeast Asia began to show a tendency to divide into two subregions, communist-controlled Indochina and the anti-communist ASEAN group.

The Bali Summit

With the end of the thirty-year war and peace achieved for the first time since independence, North Vietnam began to move towards reunification with the South. In November 1975, representatives from the North and from the revolutionary government of the South met together and adopted a programme leading to unified elections six months later and formal unification subsequently. The North Vietnamese army, which emerged triumphant from the long war, was by far the largest armed force in the region. Hardened by the combat experience against France and the United States, and strengthened by the acquisition of massive amounts of military equipment left behind by the Americans, it was indeed a threatening presence to neighbouring countries, especially Thailand. Rumours that the North Vietnamese could, if they wanted to, take Bangkok in under two weeks revived fears of the 'domino' effect and caused frantic buying of dollars on the Bangkok markets.

The prospect of such alarming developments, and the grave threat that they implied, gave a new impetus to ASEAN, which had been somewhat inactive up to then. The member countries now began to meet more often. On 17 April 1975, before the final fall of Saigon, the Filipino foreign minister, Carlos Romulo, suggested a separate meeting of the heads of the five governments, and the proposal was accepted by the Foreign Ministers' Meeting the following month. This meeting recognized the *fait accompli* in Indochina and decided to maintain a common stance as regards future developments. As a result, their recognition of the new government in Kampuchea was more or less coordinated. In June, the Philippines recognized China, and Thailand followed suit in July. Thailand also hastily began negotiations for the reduction or withdrawal of the American bases on its territory.

The leaders of the ASEAN countries were active in visiting the other member countries: in July 1975 the Thai prime minister, Kukrit Pramoj, visited all the other four ASEAN capitals; Datuk Hussein Onn, who became Malaysian prime minister on the death of Tun Razak in January 1976, visited Singapore, Jakarta and Bangkok almost immediately on taking office. The fall of Saigon provided ASEAN with the impetus which brought into play the results of eight years of accumulated effort in communication and understanding, and helped to replace deeply rooted nationalistic thinking and patterns of behaviour with the concept of regionalism.

*   *   *

It was not so much that the ASEAN leaders were worried about a direct attack being made in force by the regular Vietnamese army; what they feared, rather, was Vietnamese support for the communist movements and insurgencies inside their countries. Each country, well aware of the past twenty years of the region's history, appreciated that dangers of this kind could not be prevented by guarantees or military agreements with the great powers. It was clear that the only avenue open for counteracting such threats was that of social and economic development and political stability. In this way the essentially Indonesian concept of 'resilience' came to be accepted by the whole of ASEAN in a regional as well as a national context.

Now that their goals were more clearly defined, officials from the ASEAN countries were constantly in touch with each other, and held numerous informal meetings in addition to those that were regularly scheduled (such as the November 1975 Economic Ministers' Meeting and the February 1976 Foreign Ministers' Meeting). This paved the way for the Association's very first summit meeting, which was held in Bali on 23-24 February 1976. The results of the two days of discussions among the five leaders are seen in two documents, the Declaration of ASEAN Concord and the Treaty of Amity and Cooperation. These two documents gave expression to the perceptions, philosophies and strategies of the five countries and were of great importance in deciding the future political framework of Southeast Asia.

The Declaration of Concord has as its objective the pursuit of political stability, and it therefore speaks of taking 'necessary steps' (1) to settle a procedure for the peaceful resolution of intra-regional disputes and (2) to secure the realization of the ZOPFAN concept. There were many reasons, arising from the very diversity of the five countries, for regional disputes, such as the Filipino claim to Sabah, the Thai-Malaysian border problems and the Muslim uprisings in Mindanao in the Philippines, to occur. Moreover, since disputes of this kind easily invited interference by countries outside the region, particularly the great powers, it was very important for the nations in the region to be able themselves to contain them if they were to prevent superpower intervention.

The Treaty of Amity and Cooperation laid down that the five countries should 'at all times settle such disputes among themselves through friendly negotiations' (Article 13), and that in the event of no solution being achieved through direct negotiations, a High Council of ministerial rank should be set up to 'recommend to the parties in dispute appropriate means of settlement . . . [and] when deemed necessary . . . recommend appropriate measures for the prevention of a deterioration of the dispute' (Article 15).[1] These articles of the Treaty had no compulsory provisions, but they demonstrated quite clearly the political will of the five countries to try to contain regional disputes, and indeed since that date all regional disputes have been dealt with within the region itself.

According to a confidential position paper prepared by the ASEAN Secretariat at the time, the procedure for implementing the ZOPFAN concept — the prime political goal of the Declaration — would involve (1) the establishment of a mechanism for settling regional disputes, (2) a declaration by the United Nations of a ZOPFAN in Southeast Asia, (3) the establishment of Southeast Asia as a nuclear-free area, and (4) the conclusion of non-aggression treaties with external powers. This position paper reflects the preoccupations of ASEAN at the time and constitutes the logical basis for the birth of these two major documents.[2]

In the economic field, the Declaration of Concord provides for the mutual provision of food and energy in times of emergency and, following the guidelines of the 1972 UN experts' report, designates as longer-term objectives cooperation in large-scale industrial projects and preferential tariff agreements. It was also decided that a meeting of the Economic Ministers should take place without delay to coordinate and implement these proposals.

Despite all the careful consultation that had preceded the summit, the discussions at Bali faced numerous difficulties because of the varying levels of economic development and the divergences in perception of the five countries. The Philippines mediated between an economically radical Singapore and a cautious Indonesia by suggesting dropping the idea of creating a free trade area and concentrating instead on the tariff agreements and industrial projects referred to in the Declaration, while it took the personal intervention of President Suharto to secure the signature of the Treaty of Amity and Cooperation when Malaysia showed resistance until the very end on account of the Sabah issue. Nevertheless, as one senior ASEAN official has said, the fact that a solution to these difficulties could be found through discussion among the heads of state themselves is a tribute to ASEAN's particular way of working — a pattern of consultation that laid the foundations of its future success.[3]

Security aspects receive little mention in the two documents, except for a short reference in the Declaration of Concord calling for the 'continuation of cooperation on a non-ASEAN basis between the member states in security matters in accordance with their mutual needs and interests'.[4] ASEAN's intention was to try to avoid becoming a military alliance while at the same time taking into consideration the rivalry with the newly unified Vietnam. Article 18 of the Treaty of Amity and Cooperation lays down that ASEAN should be 'open to accession by other states in Southeast Asia', so leaving open the possibility that Burma or even Vietnam might join at a later date.

Prime Minister Lee Kuan Yew, in his opening speech at the summit, referred to the EEC and Comecon, arguing that the EEC's formation had been accelerated by competition from Comecon, and said that ASEAN, too, was forced into competition with the three Indochinese countries. China acknowledged the new attitude which was forged at Bali, and in May 1976, when Prime Minister Lee was visiting Beijing, the Chinese leadership expressed approval of ASEAN and urged that the 'necessary steps' be taken to realize the ZOPFAN concept. As relations with Vietnam began to deteriorate, China apparently felt it necessary to cultivate friendship with ASEAN. Moreover, at a period when the Soviet Union was endeavouring to move closer to Southeast Asia through its collective security proposal, the ZOPFAN concept could be seen by China as fitting in well with its anti-hegemony strategy, at least in the sense that ZOPFAN could act as a breakwater against the Russian tide. Again, when President Marcos visited Moscow in June 1976, Brezhnev was at least not negative about ASEAN; perhaps the Soviet leadership appreciated that, while ASEAN's stance could not be considered pro-Soviet, as long as it served to put a brake on Chinese initiatives, it could serve the general Soviet interest.

It is significant that in this respect ASEAN's stance of 'anti-communist neutrality' agreed with the interests of the two communist giants. This only served to heighten ASEAN's confidence. Even Vietnam, the country which had been the most critical of ASEAN, sent Phan Hien, the deputy foreign minister, on a tour of four ASEAN countries (Thailand being the exception) soon after Vietnamese reunification had been accomplished in March 1976. He declared that he had noted that ASEAN was not a military alliance and he guaranteed that Vietnam 'would export [to ASEAN] neither arms nor revolution'.

Although Vietnam joined Laos in attacking the ZOPFAN concept at the August 1976 Colombo conference of non-aligned states, by 1977 there were signs of a softening in its attitude to ASEAN. As for the United States and the other Western powers it was natural that they should welcome the Association's self-reliant stance. Thus, with the Bali summit as the turning-point, ASEAN's international legitimacy rose rapidly, so helping to foster an environment conducive to economic growth.

The Defence Debate in Japan

The fall of Saigon had a great impact also on Northeast Asia. There was a prevailing fear that, riding on the wave of victory in Vietnam, North Korea might attempt some action in the Korean peninsula. Some credibility was given to this fear, since President Kim Il-sung visited Peking in April 1975, the very month in which Saigon fell, and it was rumoured that he might visit Moscow as well. The problem was how the United States would react. Although Kissinger had remained as Secretary of State when, in August 1974, Gerald Ford took over the presidency from Richard Nixon, thereby providing some continuity in foreign policy, American domestic politics had still not recovered from the aftermath of Watergate (in particular, Congress's control over administration had tightened) and the antiwar mood had spread throughout the country.

South Korea was therefore very apprehensive about what response the United States might make if North Korea were to take action, especially since memories of the Nixon administration's 1970 decision to withdraw troops were still fresh. Alarmed also by the American policy-makers' seeming lack of understanding of, or sympathy for, the precariousness of the security situation in Korea, the South Koreans tried to explain their position through the US mass media, by means of a series of full-page advertisements. Also, from this time on, they attempted to influence American policy through systematic approaches to and persuasion of American politicians, in what later developed into a scandal known as 'Koreagate'.[5]

However, by this time, the United States was quite aware of the possibility that the impact of its defeat in Southeast Asia might spill over on to the Korean peninsula and was determined to pre-empt the danger. In a series of public statements, the US government reiterated its guarantee of the security of South Korea as a form of warning to North Korea. In the middle of the 1975 Mayaguez incident (when a US merchant ship was detained by the new Kampuchean government), Kissinger warned North Korea 'not to make a mistake in questioning the validity of the US security commitment to South Korea'.[6]

*   *   *

Japan was slower than South Korea in realizing the security implications of America's new Asian policy following the Guam Doctrine. While South Korea was engaged in an attempt to restore the American commitment in the region, Japan was preoccupied with the effects of the Nixon shocks and the oil shock, and the government had begun slowly to move away from the foreign policy line of the United States, with the aim of widening its political options as well as increasing its economic independence by diversifying its sources of supply. However, the fall of Saigon and the subsequent uncertainty in the region caused a subtle shift in this trend. The US presence was re-evaluated in the security context, and its importance was grudgingly admitted even by the left-wing intellectuals. As though to demonstrate this change of mood in the country, there was a marked decline in opposition to the existence and use of the US military bases, particularly for operations outside Japan, which, during the war in Vietnam, had seldom failed to become targets of protest. In contrast, when these bases were actually utilized during the Mayaguez incident, there was virtually no protest at all.

It seems that the fall of Saigon drove home to the Japanese the implications of the reduced American involvement in the region, which in turn altered the nature of the defence debate. For instance, the concept of 'unarmed neutrality', which the JSP had flaunted for the past two decades, began to lose its appeal, as the public began to realize that more realistic policies were needed in order to respond to security challenges in the post-Vietnam situation.

This showed that the defence debate, which had been the most controversial issue in Japanese politics for more than two decades, was not really a debate on realistic options for a defence policy, but was a ploy on the part of the opposition to check the perceived aim of the long-reigning LDP government to turn Japan's clock back from post-war democracy to the pre-war style of autocracy and militarism. In other words, the defence issue in Japan revolved not around a fear of aggression from outside, communist or otherwise, but around the public fear that the strength and legitimacy of its 'democracy' — which, after all, had been imposed upon the country by the Occupation — were at risk.

The JSP's concept of 'unarmed neutrality' had its roots in the party's 'four principles of peace' (propounded in 1950), which called, among other things, for permanent neutrality and no rearmament. As a necessary condition to make such an option viable, 'total peace' was required — meaning a peace treaty with all the former Allied powers, including the Soviet Union, as against 'partial peace', which meant a treaty with only the United States and its current allies (as conceived by the United States at that time).[7] In the early 1950s, this line of argument had considerable appeal for a people whose pacifist inclination had been threatened by the prospect of a cold-war confrontation which might involve Japan in fighting the 'monolith' of the Soviet Union and China, merely by associating itself with the United States. Extremely wary of the notion of having to defend their country in such a precarious situation, which would inevitably entail a substantial military build-up, the Japanese were hoping against hope that they could somehow obtain a joint non-aggression agreement from all the great powers, so as to make possible something approaching a 'zone of peace, freedom and neutrality — Northeast Asian version'.

During the 1950s, such arguments, though obviously too idealistic to be a practical policy option, were nevertheless quite effective in winning support away from the LDP and, by implication, from the United States. However, as Japan entered the 1960s, the JSP and other left-wing parties began to fare less well. For one thing, the LDP had learnt, through the experience of confrontational politics, how to pre-empt the thrust of the opposition by presenting its policies in a much more sophisticated and palatable form than before, as shown in its non-ideological stance and well-calculated 'economistic' policies. (Also, rapid economic growth had rendered the concept of class struggle less meaningful and somewhat out-of-date.) For another, the Sino-Soviet rift was a blow to the concept of 'unarmed neutrality', while the fiasco of the Cultural Revolution undermined the credibility of the leftist viewpoint in general. (Since the early 1950s, pacifism in Japan had been more or less identified with the aspirations of the proletarian masses, while rearmament was associated with the capitalist elite, which was believed to favour it as a means to amass wealth and power.) Moreover, China's express support of Japan's alliance with the United States, even including military cooperation (which it viewed in the context of the global struggle against Soviet hegemony), weakened the arguments of the anti-militarists. In these circumstances, it was only natural that the fall of Saigon should accelerate an existing trend and help to restore a measure of realism to the defence debate in Japan.

Even in the first half of the 1970s, the public image of the Self-Defence Forces was low, many Japanese considering their most suitable role to be assisting in disaster relief. Moreover, the perennial controversy about their constitutionality still remained a thorny legal and political issue. The JSP, the JCP and other left-wing bodies continued to adhere not only to the line that they were unconstitutional, but tried to frustrate the government's efforts to implement various facets of security cooperation with the United States, under the treaty stipulations, such as assisting US air, sea or ground manoeuvres or allowing nuclear-powered submarines and warships to make temporary visits to various naval bases. These various issues were so difficult to deal with that the government took to trying to avoid having to make controversial decisions altogether, thus producing a kind of immobilism in Japanese politics.

In the mid-1970s, however, changes in the international environment weakened the pacifist atmosphere that had been inhibiting debate about defence needs. Under Prime Minister Miki, some attempt was made to improve the image of the Self-Defence Forces; Defence White Papers began to be issued annually from 1976 (the first-ever Defence White Paper, in 1970, had created such a storm that for six years none was issued); and in the autumn of 1976 the Cabinet approved the National Defence Programme Outline, aimed at providing Japan with a more flexible and balanced defence capability. (The ease with which a defecting Soviet pilot was able to fly his MIG-25 through Japanese air defences and land in Hokkaido in September 1976 suggested that there was room for improvement in the nation's overall security system.) On the other hand, the relatively 'dovish' Miki offset this planned enhancement of Japan's defence capability by pushing through the ratification of the Nuclear Non-proliferation Treaty (originally signed in 1970), broadening the definitions used in the ban on weapon exports first introduced by Sato in 1967, and officially endorsing what had become standard practice, namely that the defence budget should be limited to a maximum of one per cent of GNP.

In retrospect, the relative immobilism of the earlier Japanese governments is understandable, given that the anti-militarist rhetoric of the left-wing opposition contained ominous overtones of the radicalism which had effectively toppled one government and aborted the visit of an American president. In addition, these left-wing activities may have had the merit of helping the government to fend off American pressure to expand defence expenditure and to provide more positive cooperation in the war in Vietnam. It would have brought home to the US administration, as mentioned in the preceding chapter, the danger of the left-wing opposition getting out of hand and threatening not only Japanese political stability but also the basis of the Japan-US alliance.

At a time when the cold war was at its height, it was natural that each superpower should be more interested in those countries that seemed to be wavering between the two sides than in those that were already committed, and should endeavour to avoid pushing them into the opposing camp and, if possible, lure them to its own side. Thus, the United States was very tolerant of Japan's burgeoning trade relations with China, North Korea, North Vietnam and East Germany, with which Japan did not even have diplomatic relations. In fact Japan was unique in that, without adopting an attitude of non-alignment or neutrality, and with the government clearly pro-American, it was able to maintain a virtual freedom to trade across the Iron Curtain.

Perhaps this was made possible because a similar mechanism was at work in the communist countries, which were also eager not to alienate Japan completely. The sight, from time to time, of left-wing demonstrators surrounding the Diet building, keeping the conservative government in a state of siege, as it were, may have suggested to the communists the possibility, however remote, of Japan one day dislodging itself from the US camp. Before the normalization of relations, China used to talk about its friendship with the 'people' of Japan and tried to support the anti-government forces through various 'private' channels, while the Soviet Union also tried, though somewhat clumsily, to draw Japan closer, using natural resource developments in Siberia as a bait.

It would certainly be going too far to suggest that there was some kind of collusion between the government and the opposition to work covertly together in order to provide Japan with the best possible deal in any given circumstances. There was a long-standing enmity and distrust between these two, based on totally different ideological orientations, which would make such a partnership impossible. Yet the established parties and the opposition, including the unions, depending upon the nature of the issue, were not always so far apart as they sometimes appeared to be. For one thing, in the largely consensual political culture of Japan, the opposition could not afford to ignore the public's perception of national interests. The way the JSP and other opposition parties had cooperated with the Tanaka government in connection with the normalization with China suggests that, in the Japanese political environment, the opposition could work to the same end as the government without anything being explicitly stated. Although this kind of behaviour invited criticism as being ambiguous and lacking in principle, it may have strengthened Japan's position, because, particularly in the area of economic policies, it helped the country to choose an optimal course out of a range of diverse and ideologically contradictory options.

Japan's 'Equidistance' Diplomacy

The government of Miki Takeo, who succeeded Tanaka as prime minister in December 1974, inherited two very sensitive unsettled diplomatic problems, namely the conclusion of peace treaties with, respectively, China and the Soviet Union. Although Sino-Japanese economic links had expanded rapidly after the normalization of relations in September 1972, progress on the peace treaty had been blocked by China's insistence on the anti-hegemony clause.[8] In the case of the Soviet Union, although discussions had been held about cooperation in a number of large-scale projects, such as the development of Tyumen oil and Yakutia coal resources and the construction of a second Trans-Siberian Railway, the peace treaty was still outstanding because of inability to resolve the issue of the Northern Territories. When Prime Minister Tanaka visited Moscow in October 1973, he and Brezhnev signed a communiqué calling for further dialogue at an appropriate moment in 1974, in order to settle the questions left 'outstanding' from World War II, as well as to sign a peace treaty. The Japanese assumed that the 'outstanding questions' included the issue of the Northern Territories, but the Soviet Union denied this and so the negotiations reached an impasse.[9]

98 A New Paradigm of Relations

In mid-January 1975, in order to display an even-handed approach to Japan's two communist neighbours, the Miki Cabinet sent Hori Shigeru, a leading LDP politician, to Beijing and Foreign Minister Miyazawa Kiichi to Moscow, so as to reopen negotiations with both countries simultaneously. The government was eager to reach a speedy settlement of both negotiations, but particularly to conclude a Sino-Japanese peace treaty, which would have been very popular and could therefore strengthen its internal position, which was none too strong initially. However, in a situation where the competition in Asia between China and the Soviet Union was becoming quite fierce, the anti-hegemony clause was too important and sensitive an issue to juggle with, simply for the sake of domestic politics. As a result, no conclusion was reached in either of the treaty negotiations, and negotiations with the Soviet Union failed even to take off.

The competition for influence between the Soviet Union and China can be traced back to the 1955 Bandung Conference, the first international political conference of African and Asian states, when the Soviet Union, which was not invited, was forced to watch helplessly as China scored a historic success. It was able to get its own back, however, in the second half of the 1960s, when China lost its foothold in Indonesia by being implicated in the abortive *coup d'état* of 1965. Also, because of the turmoil of the Cultural Revolution, China was not in a position to formulate a coherent Asian policy, and the Soviet Union took advantage of the situation in order, slowly, to establish its position in Southeast Asia.

There were several reasons for the Soviet Union's latest thrust into Asia. First, its navy had grown in strength to the extent that it had become a real global navy, and so it had to find docking and repair facilities in the broad area surrounding the Eurasian continent, from the Baltic to the Sea of Japan, as well as to secure access through such strategic waterways as the Malacca Straits. Second, as its position relative to the United States rose, it began to expect the Asian countries to recognize it as a world power, with a right to have a voice in their affairs. And, third, it had become necessary to mount a determined campaign throughout the region if an unruly China were to be circumscribed. In order to realize these objectives, in January 1969 Brezhnev put forward the idea of an Asian collective security system and began to sound out the Asian countries for support.[10]

China, for its part, was beginning to view the relative positions of the United States and the Soviet Union in their struggle for power as having been reversed, because of the Vietnam war.[11] For a long time, China had regarded the United States as being on the offensive and the Soviet Union on the defensive. Subsequently, this had changed, because while the war in Vietnam was draining the United States both economically and psychologically, the Soviet Union had accumulated such enormous strength that it was able to move on to the offensive, seriously aiming for world domination. For China, the normalization of relations with the United States and Japan was conceived as a countervailing action to the unrestricted expansion of Soviet influence.

The Soviet response was to seek ways, in turn, to contain China — the Asian 'collective security system' being one avenue, as shown by the signing of a friendship treaty between India and the Soviet Union in August 1971. Also, with the Sino-American rapprochement providing the momentum, Soviet relations with Vietnam became much closer, with military assistance being greatly expanded. In these circumstances, it was natural that China should be nervous about North Korea, and even about Taiwan, which was rumoured to be covertly exploring relations with the Soviet Union. Seen from this Chinese perspective, Japan was the country that could most be taken for granted, because of its persistent emotional attachment to China as opposed to its fear and mistrust of the Soviet Union. In this sense, and also with its potential capability of extending substantial economic assistance, Japan was in a strong position in the negotiations with China for the peace treaty, although it did not try to capitalize on this advantage.[12]

In view of Japan's economic strength, many people inside and outside Japan argued that it should have responded more positively to the post-Vietnam situation, and should have tried to secure a lasting framework of peace and security in Northeast Asia. As it was, it did little beyond resisting Chinese pressure to give way on the hegemony clause, which would obviously have offended the Soviet Union, and resisting Soviet pressure to conclude a friendship treaty instead of a peace treaty (negotiations on which continued to be deadlocked on account of the dispute over the Northern Territories). In other words, Japan did no more than adhere to its 'equidistance' posture towards China and the Soviet Union. Although this stance may have helped to preserve a delicate balance in the region at a crucial period, it has been criticized as being too passive an attitude and doing little to strengthen the security of Japan's two neighbours, South Korea and Taiwan.

It was fortunate for the region that the feared invasion by North Korea did not occur, thanks partly to the USA's prompt and timely warning. Even if it had happened, though, it is probably safe to assume that neither China nor the Soviet Union would have been in a position to support the North Koreans, if only because of the deep division between themselves. However, if the North had invaded the South at that time, one wonders what in fact Japan could have done to support its allies, except, again, to provide the United States with logistic bases. Japan's position in this respect was not different from that of twenty years previously, when its four war-torn islands were still struggling on the breadline.

It has been alleged that Japan's largely self-centred attitude was due to a deliberate refusal to graduate from the comfortable position of dependence on the United States, which it found it convenient to believe would exonerate it from various international responsibilities. Also, successive LDP governments have been criticized for not making a serious attempt to alter a political system that could be paralysed in this way by the unruly rhetoric and action of the opposition, rendering the country virtually unable to formulate its own policy for the region.[13]

Although these criticsms may have some validity, one can only ask, again, what Japan could or should have done at that particular juncture. It is true that its economic strength had grown, but the oil shock made it painfully clear that, as the economy expanded, so the nation's economic vulnerability and dependency also increased. Moreover, Japan had to realize that economic strength does not put a country in a position of overall strength unless it is supported by military power and a suitable political arrangement. The fate of the programme of natural resource development in Siberia is a case in point. Of the many ideas which were floated, few materialized, primarily because the Japanese were afraid to make a substantial and long-term commitment with the Soviet Union by themselves. They sought the participation of American business interests in the proposed ventures, financially or technologically or both, since a US presence, they believed, would strengthen their bargaining position vis-à-vis the Soviet Union, but unfortunately they were unable to secure sufficiently solid backing.

A precondition for Japan to play a more active role in the tight and sensitive framework of the Northeast Asian balance of power would have been a more positive defence policy — an increase in armaments and closer cooperation with the United States. However, it is difficult to know what kind of reaction such a policy would have provoked from China and the Soviet Union. For instance, in the 1950s, if Japan had agreed to strengthen and expand its military capabilities as the United States wished, it could have so alarmed both China and the Soviet Union that the subsequent course of the Sino-Soviet rift would have been different. In certain circumstances, a decision not to take a clear political stand can be a wise one. In the case of Japan in the mid-1970s, it can be argued that its inaction, or refraining from action, supported the *status quo* in the region, and gave the government a certain, if limited, degree of flexibility.

The issue of the northern islands, which had long been the principal stumbling-block in Japanese-Soviet relations, worked, in a sense, as a blessing in disguise, for it enabled Japan to pursue its equidistance policy. Since this issue had dominated the Japanese-Soviet dialogue for so long, with a continual back and forth of arguments, Japan was able to use it as a sort of smokescreen in its dealings with the Soviet Union, behind which it could maintain some leeway to move closer to China without antagonizing the Soviet Union too much. In actual fact, Japan often tilted towards China in its diplomacy and acted in an unfriendly manner towards the Soviet Union, as shown in the September 1976 incident of the defecting Russian pilot, when the Japanese were unreceptive to Soviet demands to return the pilot and the plane at once. Japan could explain such incidents as inevitable consequences of the Russian refusal even to talk about the northern islands. (Our policy is to maintain equidistance, but you make it difficult for us to do so.) Moreover, the northern islands issue provided Japan with useful evidence of its anti-Soviet attitude to show to the United States and its allies.

Admittedly, the problem of the northern islands, like all territorial disputes, is a very difficult one to solve, and furthermore the government's claim to the islands has the full emotional backing of the Japanese people. (All the political parties, even the JCP, support the government's position on the issue.) However, the fact that there are no longer any Japanese living on the islands makes it easier to use the disputes for diplomatic purposes.[14]

The Fukuda Doctrine

Even in 1975, nearly eight years after the formation of ASEAN,

Japan's evaluation of that organization was, at best, ambivalent. There were several reasons for this. First, Japan's initial encounter, at an official level, with the ASEAN countries had been unfortunate, and had merely reinforced its suspicion that ASEAN was a collective bargaining forum of the developing countries. At a meeting in April 1973, the ASEAN foreign ministers had protested to Japan that its synthetic rubber production and export was having a damaging effect on the economies of the ASEAN rubber-producing countries, and had requested discussions. Negotiators met four times before a satisfactory solution was found, with Japan agreeing to provide various forms of assistance to the ASEAN rubber industries. These ASEAN demands coincided with the period of anti-Japanese movements in Southeast Asia (1973–4), which led Japan to adopt a rather cautious attitude towards ASEAN. In fact, in view of the generally unfriendly attitude then prevailing in the region, it was inclined to feel that even if it adopted a more cooperative approach, it could well meet with a rebuff.[15]

Second, the concept of neutrality, which ASEAN had been emphasizing since the 1971 Kuala Lumpur Declaration, was problematic for Japanese policy-makers. Despite the variety of oppositional opinion in Japan, it was indisputable that the country belonged to the Western alliance and depended completely on the American nuclear umbrella. For all the differences in semantics, as well as hidden intentions, among ASEAN's members, the Japanese government was nervous about supporting an organization which professed neutrality as its principal platform. It was also afraid that the ASEAN position on the 'temporary' nature of foreign military bases might stimulate further debate in Japan about abolishing the Japan-US Security Treaty.

Third, Japan was more used to interacting with other countries, including the ASEAN member states, on a bilateral basis. This was particularly true of those aspects of its diplomacy which related to its effort to ensure the supply of various natural resources. Moreover, the organizational structure of its Foreign Ministry did not make it easy to cope with an international organization in which Japan was not itself a participant.[16]

However, the fall of Saigon and the ASEAN summit conference in Bali changed Japan's attitude and brought a new element into its policy considerations. It was clear that if the fall of Saigon were to bring into play the 'domino' effect, and, if, as a result, Southeast Asia were to come under Vietnamese domination, then the base of Japan's economic security would be seriously jeopardized. It was therefore more than welcome to the government when the ASEAN nations widened their contacts and strengthened their cohesion after the fall of Saigon, and began to take positive steps to show that they had the political will, as well as a programme, to preserve the stability of the region. It goes without saying that ASEAN's emphasis on socio-economic growth and development as the basis for national as well as regional resilience coincided nicely with Japan's own way of thinking.

At that period, increasing international criticism was being levelled at Japan for not taking on responsibilities commensurate with its economic strength. Japan had also been subject to criticism from the OECD for the low level of its developmental assistance. The United States voiced complaints that, while its national strength was being sapped in Vietnam, Japan had done nothing except amass its wealth, contributing little to regional stability. It would therefore be very timely if, in the wake of the American withdrawal from the region, Japan were to take up the challenge of giving concrete support to efforts at regional organization and display its willingness to take international responsibility.

This obligation was recognized by Japanese Foreign Ministry officials at a meeting held in Hong Kong immediately after the Bali summit.[17] The Japanese Ambassador to Singapore, on behalf of his country, made the statement — unusually positive for his government thus far — that 'Japan is likely to step up its economic role in ASEAN, as the five-nation group is most important in Japan's foreign policy towards the region'. He added that 'Japan would do whatever was possible to promote peaceful coexistence between ASEAN and Indochina'.[18] This suggested that Japan's future policy towards ASEAN would have two components: to help strengthen the resilience of the ASEAN countries as a group, and to seek stability and prosperity for the whole of Southeast Asia, including Indochina.

After the fall of Saigon, Southeast Asia had been divided into two blocs, the communist countries, centred on Vietnam, and the anti-communist countries, most of which closed ranks in ASEAN. Japan did not favour the idea of adopting an anti-communist position by propping up ASEAN against Indochina, and wanted to encourage peaceful coexistence on the part of both blocs — a policy that it had followed ever since the end of World War II. Moreover, the Foreign Ministry was aware of the difficulty it would have in rallying domestic support, particularly from various other government agencies, and knew that it would be better placed if the basic rationale of the proposed aid package to ASEAN were presented, not as following an anti-communist line, but as a means of encouraging self-reliance on the part of the recipients, and hence regional stability.

As already noted, the ideological dichotomy in post-war Japanese politics was so marked that few policies could expect to enjoy general and consistent support. Thus, in order to implement a positive policy towards ASEAN, it was essential to try to generate a broad consensus within the government, both on the importance of ASEAN and on the Foreign Ministry's proposed policies towards it. Without such a consensus, various ASEAN-oriented priorities, such as economic assistance and trade concessions (which inevitably conflicted with domestic interests), would get pushed around by pressure groups. The Foreign Ministry had had a number of bitter experiences of this kind, often bringing upon itself criticism from outside that there was a difference between what Japan said and what it did.

So, after the Bali summit, the government undertook a large-scale *nemawashi* (laying the groundwork) operation — a moulding of opinion, both external and internal. With regard to ASEAN, the Foreign Ministry invited to Japan, in turn, the secretary-generals of the five national secretariats and also the head of the newly formed ASEAN Secretariat in Jakarta. The purpose was partly to ascertain that there was no dissatisfaction, let alone opposition, in ASEAN about the general trend of Japanese thinking, and partly to try to establish a pattern of working with ASEAN as a multinational organization. This entailed reaching an understanding on how to differentiate such multilateral relations from the traditional bilateral links that Japan already had with each of the ASEAN states. Internally, an agreed policy was worked out with the other ministries concerned — the Ministry of Finance, MITI, the Ministry of Agriculture, Forestry and Fisheries and the Economic Planning Agency. This led to the formation of the Japan-ASEAN Forum as a venue for the discussion of future Japan-ASEAN relations, with the first meeting being held in Jakarta in March 1977.[19]

ASEAN, too, began at this time to give serious thought to the question of relations with Japan, having become aware that these relations could hold great importance for its future. Although the 1973-4 anti-Japanese sentiment had not completely disappeared, Japanese businesses had made efforts to respond more sensitively to the requirements of local populations, while the cultural exchange programmes promoted by the Japan Foundation and other private organizations were slowly helping to increase mutual understanding.

However, the main reason for ASEAN's new attitude was a recognition of the potential importance of Japan in the post-Vietnam situation in Southeast Asia. When, after the Bali summit, the outlines of ASEAN's future development, based on national and regional resilience, began to emerge, the importance of Japan's active involvement suddenly became apparent. Also, Japan's economic performance — the first major economy to return to the path of growth after the 1973 oil shock — aroused the ASEAN countries' admiration.

* * *

In August 1977, following this growth in mutual understanding and the *nemawashi*, Prime Minister Fukuda set out on his tour of ASEAN countries, first attending the ASEAN summit meeting in Kuala Lumpur, and then visiting each of the five member states and Burma. This tour was of crucial importance, not only because an unprecedented amount of economic assistance was promised to these countries — a total of Y407 billion (US $1.55 billion), of which Y268 billion ($1 billion) would be allocated to the five ASEAN Industrial Projects (AIPs)* — but also because it established a framework of relations between Japan and Southeast Asia.

In the last capital he visited, Manila, on 18 August, Prime Minister Fukuda, before an invited audience which included President Marcos, made a speech in which he stressed three particular points: that Japan would not become a military power; that its economic links with the ASEAN countries were founded on a 'heart-to-heart' relationship, in which there was an equal partnership among fellow Asians; and that it would try to foster a 'relationship based on mutual understanding' between ASEAN and the Indochinese countries. His speech was received with enthusiastic applause, and, in response, President Marcos said that he admired Fukuda for his efforts to move Japanese foreign policy in a new direction and congratulated him on his leadership, adding, 'We have been waiting a long time for this kind of attitude to appear in Japan. Now, without any hesitation, I can say

---

*This was a scheme whereby a large industrial plant would be set up in one of the five countries, with a view to producing the region's total requirements for that particular product. Each ASEAN member was to have one such large plant for which it would submit a proposal, and each plant would be owned jointly by all five. Three proposals for AIPs have now been accepted by the ASEAN economic ministers — in Indonesia, Malaysia and Thailand — and are being implemented.[20]

that ASEAN really has found a true friend in Prime Minister Fukuda.'[21]

Certainly the position adopted by Fukuda — the Fukuda Doctrine, as it came to be called — was a new departure. The recognition of a special relationship with ASEAN was a conspicuous shift from the omnidirectional stance previously adopted — a policy based on the need for a global network of markets and supply of natural resources. Also, the use of such expressions as 'heart-to-heart', which are not found in normal diplomatic jargon, appealed to people's emotions and had an even profounder effect than the commitment to economic assistance.

Japan's advocacy of a relationship based on mutual understanding with the three Indochinese countries (made despite the reported objections of some leaders in the region, including Thailand's Prime Minister Thanin Kraivichien) was one of the rare occasions on which it showed a positive interest in the political framework of Southeast Asia as a whole. That a country like Japan could take such a stand perhaps suggests that Southeast Asia, despite being divided into the communist and anti-communist camps, had, in the global sense, ceased to be a cold-war battleground. Only in a situation in which the countries of the region were able to develop their own systems, according to their own aspirations, without great-power interference, would a country like Japan — which from the beginning had been extremely wary of being involved in the cold war — be prepared to take its own line.

The involvement of the ASEAN countries in the preparatory stages of the Fukuda visit was quite remarkable. It was as if these countries made a concerted effort to make the impending visit of the Japanese prime minister as productive as possible. At various stages, senior ASEAN officials and ministers, such as the Indonesian minister of economic affairs, Wijoyo Nitisastro, visited Japan and presented their cases so as to be able to influence Japanese thinking. In May 1977, President Marcos also visited Japan. These contacts, as well as the speeches and statements which the leaders of the region had made in their own countries, were orchestrated for the purpose of pushing and prodding the process of Japanese policy formation and decision-making. By these means the ASEAN policy-makers involved themselves informally in the Japanese decision-making system, and in the process they came to understand more about that system and how to obtain the responses that they wanted. In this way, through direct involvement in Japan's *nemawashi* process, the

ASEAN countries came to see that there were certain similarities between their own political cultures and that of Japan, in that — on both sides — decision-makers needed to establish a consensus before they could put their policies into effect. The success of this whole episode in Japan-ASEAN relations must be attributed to the fact that it was a truly joint effort — a fact that may suggest the emergence of a *modus operandi* peculiar to Asia. Certainly, the rapport which developed between Japan and ASEAN as a result of this cross-fertilization was to constitute one of the concrete sources of stability for Southeast Asia after the fall of Saigon.

The Japanese policy-makers were pleased that there was little ideologically motivated internal opposition to the proposal for cooperation with ASEAN. Several reasons for this can be suggested: (1) ASEAN was a group of developing countries; (2) ASEAN stood for neutrality, with three of its members belonging to the non-aligned movement (NAM); (3) neither left nor right had reason to dispute the truth of ASEAN's determination to become more self-reliant; (4) most of the communist countries — namely the Soviet Union, China and, to a certain extent, even Vietnam — seemed to take a positive attitude towards ASEAN; and (5) the absence of ideological dispute, such as has marked relationships with, say, South Korea, was clearly an important ingredient in the success of the enterprise.

Helped by this favourable environment, the Japanese public has begun to understand what ASEAN is about. The unanimously enthusiastic welcome for Fukuda on his trip round the ASEAN countries, in contrast with the violent demonstrations that greeted Tanaka in 1974, naturally accelerated the process. Starting from the Fukuda visit, the Japan-ASEAN relationship has gradually grown closer and has indeed become endowed with the qualities of a 'special relationship'.

* * *

The presence of the Prime Ministers of Australia and New Zealand in Kuala Lumpur has come in for relatively little notice because of the high visibility of Prime Minister Fukuda. Nevertheless, the fact that three prime ministers attended the ASEAN summit together was significant in that it demonstrated the importance which ASEAN had acquired internationally as a regional organization. It was a far cry from ten years earlier, when even its formation, in August 1967, drew little attention from the outside world. In the cold-war environment of the time, it was generally assumed that it would be difficult for a group of developing countries, with relatively little political influence, to launch a scheme of this kind on their own. Furthermore, the differences among the member countries — in size, ethnic composition, religion and culture — were regarded as being an obstacle to the achievement of any kind of regional cohesion. Finally, the largely resource-based economies of these countries (with the exception of Singapore) were regarded as making them competitive with each other rather than complementary, and so hindering the cooperative effort. Even the members themselves were not wholly convinced of the long-term prospect for the group, although each of them had specific reasons for welcoming its formation.

Also, unlike the European Community (another successful postwar effort at regional cooperation), whose members have a largely homogeneous political orientation, ASEAN members maintain a multiplicity of affiliations, cutting across all kinds of dividing lines, ideological, religious and political. Thus, Indonesia and Malaysia are members of the Organization of Islamic Countries; Indonesia is a member of OPEC; in 1979, the Philippines hosted a summit conference of the Group of 77; Indonesia, Malaysia and Singapore still retain their membership of the non-aligned movement, whereas Malaysia and Singapore are members of the British Commonwealth as well as of the Five-Power Defence Arrangement; and finally, as a signatory of the Manila Pact and also as a frontline state facing the possible spillover of the wars in Indochina, Thailand receives special attention from the United States and its allies all the world over.

This diversity has worked to ASEAN's benefit. For one thing, it has spared it from the left-wing charge that it is a tool of American imperialism. For another, its pragmatic emphasis on indigenous socio-economic development has drawn support from the developed countries, particularly Japan. Finally, it sees to it that its members can collectively benefit from the diverse orientations of the other member countries. In other words, an outsider negotiating with one member has to keep in mind the entire spectrum of affiliations which the group as a whole enjoys. This is an ingenious arrangement, designed to enhance the bargaining power of the group as a whole as well as that of the individual members.[22] The efficacy of such group diplomacy has been dramatically illustrated by the steady increase in support that ASEAN has been able to achieve in its efforts — from 1979 onwards — to defeat Soviet attempts to oust the representatives of Democratic Kampuchea from the United Nations.*

ASEAN was fortunate to have had a group of capable leaders who, unlike some of their predecessors, knew that the legitimacy of their respective governments would be judged by their own record of internal development, rather than by their ability to make spectacular foreign-policy moves or to obtain aid and assistance from foreign powers. They were supported by a growing number of extremely able technocrats, who were eager to join hands in the national effort. These men may not have been devotees of free trade, *per se*, but they shared a forward-looking attitude to the economy, and tried to accelerate growth by making their industries more competitive, rather than protecting the weaker sectors.

ASEAN was further fortunate in that its period of growth as an organization coincided with the less confrontational political environment of the 1970s. The fact that China had noticeably toned down its ideological fervour and had begun to concentrate on the practical goal of modernization had a far-reaching effect both on ASEAN and on the region: by reducing the credibility of communism and class struggle as avenues to socio-economic development, it reinforced the legitimacy of the market-oriented policies of ASEAN governments. This shift on China's part saved the ASEAN governments, to a large extent, from having to cope with ideologically motivated internal unrest, and enabled them to concentrate on the priority of economic development. Given the fact that in many other parts of the world developing countries are still torn by ideological conflicts, ASEAN's circumstances in this respect were extraordinarily favourable to its goals.

---

*In October 1982, 90 countries voted in the UN General Assembly to support ASEAN's move to retain the UN seat for Democratic Kampuchea — 13 votes more than had been obtained in 1981. This vote effectively denied, once again, international recognition for the Heng Samrin regime, which the Vietnamese had been pushing for the past three years with the full backing of the Soviet Union and its allies. As though to cap its victory with one more prize, the Assembly also awarded ASEAN 105 votes, compared with 100 in 1981, in the voting on a motion calling on Vietnam to withdraw its troops from Kampuchea and allow Kampucheans to choose their own government.

Chapter Six
PERSISTENT CONFLICTS AND COMMUNISM IN ASIA

Indochina

Eight years after the end of the war in Vietnam, Indochina is still plagued by persistent armed conflicts and fundamental political uncertainty. There is little sign of a let-up in the competition for influence between China and the Soviet Union, with the ominous possibility of renewed interfererence in the region on the part of outside powers. Also, this is the one area in the region where the economy remains in a permanent state of stagnation. The estimated GNP of Vietnam in 1981 was a meagre $9 billion, only one-quarter of the $39.7 billion of Thailand, and one-twentieth of the ASEAN total of $192 billion, in the same year. After thirty long years of struggling for independence in two major wars, the people of Vietnam now have a living standard which has actually deteriorated since 1975, the year of victory.

The difficulties which faced the Vietnamese leadership after the war were manifold. First, during the war years, Vietnam's economy, in both the North and the South, had been sustained largely by foreign assistance. (In North Vietnam, aid from communist countries accounted for half of its GNP.) The government's main task was to distribute the aid, and therefore it had little opportunity to acquire experience in managing its own economy. Second, carried away by the glow of victory, the party leadership was too hasty in applying the communist doctrines to the war-torn economy. Third, the drying up of foreign currency after the war, owing to shrinking foreign aid, halted imports of industrial raw materials and consumer goods, which stagnated economic activity in general and clamped down on the population's living standard in particular. Fourth, the integration of the South turned out to be far more difficult than expected, its people adamantly refusing to fall in with the socialist system. Finally, aid from China was cut off in 1978, as relations between the two countries worsened (although there was some increase in Soviet aid after the conclusion of the Soviet-Vietnamese Treaty of Friendship and Cooperation in November 1978); and, after the invasion of Kampuchea at the turn of the year, virtually all the aid from Western countries and Japan was cut off.

In view of the gravity of the situation, the Vietnamese government decided, in 1979, to test out a liberalization of the economy, starting with the introduction of a contractual system in the agricultural sector. The function of state-operated markets was gradually phased out, and was replaced by a free-market system. In addition, the wages of civil servants and industrial workers were increased. Tran Phuong, the newly appointed vice-chairman of the Council of Ministers, explained the new economic policy as follows: 'When we were busy fighting aggression we could accept nearly equal incomes. Now we must motivate people to produce. If all are equally rewarded, nobody wants to work any more.'[1]

A group of Western journalists who visited Hanoi in February 1983 were impressed by the improvement in economic conditions and general living standards in the past two years, and similar reports have come from Japanese journalists.[2] More food is available in the shops and, stimulated by the liberalization policy, some sectors of industry, such as cement, electrical goods, and silk and cotton weaving, have regained some momentum. Nothing has been done, however, about many basic problems (such as the shortage of oil), so that much general industrial activity is still stagnating. Foreign currency reserves have decreased to such an extent that, at the end of 1982, they were estimated at a mere $10 million. Several Western countries have extended the repayment periods of their loans (including Japan, with a loan of 3.5 billion yen).[3] Debt-servicing in 1982 came to $284 million, which was 70 per cent of Vietnam's total exports (and 218 per cent of its exports to the West).

On the political front, Vietnam's most difficult problem has been the worsening of relations with China and Kampuchea. Even before the fall of Saigon in April 1975, relations with China were showing signs of strain. The Chinese suspected that, after the war, Vietnam would try to extend its overall domination of Indochina, which in turn would be exploited by the Soviet Union as a part of its programme to encircle China's southern flank. When Le Duan visited Beijing in September 1975 to thank the Chinese for their assistance during the war, the Chinese demanded that Vietnam accept the 'anti-hegemony' (anti-Soviet) line. When Vietnam refused, the Chinese hinted that they would have to reduce economic assistance and demanded the withdrawal of Vietnamese troops from the Spratly islands, which were claimed by both countries.[4]

Vietnam's fear of China's expansionism, which was prone to extend itself south of its borders, dated far back (to long before the arrival of Western colonialism in Southeast Asia), while Kampuchea, in turn, was afraid of Vietnam's expansionism, which was a constant threat to its eastern border.[5] In August 1975, capitalizing on these legacies of fear, China managed to get the Khmer Rouge to sign a joint communiqué opposing Vietnam's 'hegemony'. Relations between Vietnam and Kampuchea deteriorated steadily from then on, and in 1977 Kampuchea broke off official relations and expelled pro-Vietnamese elements from the Kampuchean Communist Party and from the government. China supported this move by stirring up the ethnic Chinese population in Vietnam, as well as by encouraging the Khmer Rouge to make attacks in the border areas.

The Vietnamese realized that, if they were to avoid being squeezed by these two countries, they would have to neutralize the Khmer Rouge led by Pol Pot, by force, if necessary. They assumed that, given the extreme unpopularity of the Pol Pot regime — both inside and outside the country — on account of the wholesale terrorism which it had conducted since its accession to power, they could perhaps get away with using a measure of force against Kampuchea and still maintain international support. However, in order to avoid the risk of China's retaliation, they joined Comecon in June 1978, and in November concluded their Treaty of Friendship and Cooperation with the Soviet Union, which stated that if either country were to be 'attacked or threatened with attack' the two countries would 'immediately consult with each other with a view to eliminating that threat'.[6]

In addition, the Vietnamese government tried to improve its relations with the ASEAN countries by sending Prime Minister Pham Van Dong, in October 1978, to the five ASEAN capitals. Up till then, Vietnam had ignored ASEAN, dismissing it as yet another tool of the United States and confidently preaching the ultimate victory of its own kind of revolution in Southeast Asia. This time, however, Pham Van Dong was modest and amicable, promising ASEAN that Vietnam would never interfere in the internal affairs of other countries or give support to those internal insurgencies which had long been ASEAN's principal worry.

In the following month, Deng Xiaoping visited Thailand, Malaysia and Singapore, in an obvious attempt to compete with Vietnam for ASEAN's favour, which gave an unexpected boost to ASEAN's international standing. In contrast with Pham Van Dong, he stated that government-to-government relations were different from party-to-party relations, and that China would continue to extend support to the underground communist movements in Southeast Asia, if only to prevent them from moving over to Vietnam or the Soviet Union for protection and guidance. As a result, his welcome was somewhat reserved, especially in Malaysia and Singapore. Later on, however, when Vietnam invaded Kampuchea, Pham Van Dong's friendliness was condemned as trickery, designed to conceal his real intention, while Deng recovered some grudging respect for his 'honesty'.

The Vietnamese invasion of Kampuchea, launched on 25 December 1978, was a brilliant military success, so that by 7 January 1979 Phnom Penh had been taken, and a week later Vietnamese troops had already reached the Thai border. In Phnom Penh, Heng Samrin, who had been a divisional commander of the Khmer Rouge, was installed as head of a pro-Vietnamese government, while the Khmer Rouge were relegated to fighting in guerrilla groups in the jungles. Vietnam had thus managed to avoid the danger of undergoing a pincer attack from China and Kampuchea — but the costs that it had to pay for this venture were high.

First, being condemned as an aggressor, Vietnam lost its international credibility, a loss whose effects were far more serious than it expected. (In contrast, ASEAN greatly improved its international standing, merely by opposing the Vietnamese action, as was shown by the votes that it won in successive years at the United Nations on the question of Kampuchean representation.) Second, in addition to the direct military expenditure incurred by the war, the drying up of economic assistance from abroad, in retaliation for its action, worsened Vietnam's already ailing economy. Third, China's punitive strike against its northern border in February 1979 was a blow to the credibility of communism, and by implication to that of Vietnam. Also, in the context of its international image, Vietnam again had to suffer in that, although China was criticized for its actions, on the whole it was exonerated as an aggressor punishing an earlier aggressor, while Vietnam, as the initial aggressor, received almost no sympathy.[7]

Finally, the refugee problem was intensified by the war. The outflow of ethnic Chinese from Vietnam, which was already on the rise as a result of the worsening relationship with China, increased dramatically after the invasion, making the estimated total for 1979 more than 300,000, excluding the Kampuchean refugees moving into Thailand.[8] The wave after wave of refugees arriving by boat in every area of Southeast Asia became a problem of global dimensions, although it affected the ASEAN countries most. As Singapore's foreign minister, S. Rajaratnam, put it, 'Each junkload of men, women and children sent to our shores is a bomb to destabilize, disrupt, and cause turmoil and dissension in ASEAN states. This is a preliminary invasion to pave the way for the final invasion.'[9] Mobilizing every organizational instrument at its command, such as the Foreign Ministers' Meetings and the Standing Committees, ASEAN endeavoured on one hand to set up regional reception centres to process the incoming refugees, and, on the other, to appeal for the attention and action of both the expelling country (Vietnam) and the prospective resettlement countries (the West) and aid donors.

At the seven-power summit conference held in Tokyo in June 1979, the ceiling for the participating countries to resettle refugees was raised; and this was followed by the Geneva Conference in July 1979, convened personally by the UN Secretary-General, Kurt Waldheim, and attended by leading representatives from 66 countries. The Vietnamese delegate, Phan Hien, in his country's first admission of responsibility for the situation, stated that 'Vietnam was aware of the difficulties caused by the "emigration" from Vietnam'. Waldheim disclosed that the Vietnamese government had given him a pledge to make every effort to stop illegal departures. It appeared that ASEAN's endeavour, especially through its preliminary talks with Vietnam in Jakarta, had helped to convince the Vietnamese of the immense political cost involved in the refugee problem. As a result, the conference found itself able to increase the offers for resettlement from 125,000 to 260,000 refugees and to obtain a financial commitment of $190 million, the bulk of which was pledged by Japan.[10]

Japan was predictably reluctant to make any significant increase in its rate of acceptance of refugees, because of its unusually homogeneous social structure. (Even after a 1981 upward revision, the settlement quota was only 3,000.) Rather, it opted to play a role by making financial contributions. Following its substantial pledge at Geneva in 1979, it offered a further $100 million at the second Geneva Conference in May 1980 — out of a total of $170 million requested. In addition, it offered $270 million in 1980 to help Thailand with the inflow of Kampuchean refugees at its eastern border.

Japan has been consistent in its wish to promote peaceful coexistence between Indochina and ASEAN, as expressed in Fukuda's speech in Manila in 1977. Even prior to that speech, it had endeavoured to assist in the economic rehabilitation of the reunified Vietnam, offering aid and economic cooperation. In September 1976 it offered a grant of 5 billion yen, and in October it became the first country to sign a non-governmental trade agreement with Vietnam.[11] In April 1978, it concluded an extensive cooperation agreement, in which it extended to Vietnam a loan of $100 million. Also, in December 1978, on the occasion of the visit of Foreign Minister Nguyen Duy Trinh, it offered a further package of aid which, however, was suspended because of the subsequent Vietnamese invasion of Kampuchea. Current Japanese policy can be summarized as: (1) to insist, as an immediate step, on a cease-fire in Kampuchea and the withdrawal of Vietnamese forces; (2) to coordinate its policy closely with that of ASEAN and to withhold giving aid to Vietnam at least for the present; and (3) to aim, as a long-range objective, at securing a comprehensive political solution.

* * *

Unfortunately, for a variety of reasons, the prospects for a political solution are not bright. First, the Vietnamese, having acted with such determination in 1978, are unlikely to agree to go back to the original pre-invasion situation unless they are assured of very favourable conditions. Their economy is recovering, even if only slightly, despite the burden of this war, and the assumption that, if the war continues, they will sooner or later have to compromise seems to be unjustified. Rather, their calculation appears to be that if they maintain their present position and manage to consolidate Kampuchea under the Heng Samrin government, then China may one day shift its course — influenced, perhaps, by developments in its relationship with the Soviet Union.[12]

Second, although the ASEAN member states are unanimous in opposing Vietnam's violation of Kampuchean sovereignty, they nevertheless display subtle differences of attitude towards Vietnam. Indonesia, for example, has long felt an affinity with Vietnam, largely because of the shared experience of the struggle against colonialism. This, and its long-standing antagonism towards China, conditions Indonesia — and Malaysia too — to favour a strong and independent Indochina as a buffer between China and ASEAN.[13] In fact, both Indonesia and Malaysia have made several attempts to establish an understanding with Vietnam, but nothing has come of them so far, apparently because of the inflexible attitude of Vietnam.

In addition to this diversity of opinion within ASEAN, its determined policy as regards Kampuchea's recently established coalition government has not greatly advanced matters. In July 1982, as a result of considerable pressure from ASEAN, a tripartite coalition government was formed — the Coalition Government of Democratic Kampuchea (CGDK) — by Son Sann's Khmer People's National Liberation Front (KPNLF), Prince Sihanouk's Moulinaka group/and the Khmer Rouge under the nominal leadership of Khieu Samphan. This move certainly helped ASEAN at the meeting of the UN General Assembly in October to rally support for its motion urging Democratic Kampuchea's continued membership of the UN, but it does not seem to have been particularly helpful in achieving an overall political solution. The anti-Heng Samrin forces intensified their guerrilla activity, and the Vietnamese — perhaps in response — showed greater fierceness than ever in their 1983 dry-season offensive. As for the CGDK, although the three groups have joined together, they have kept their organizations separate and each has to compete for aid from China and ASEAN by showing its effectiveness on the battlefield.

Third, China, which itself undertook a punitive attack on Vietnam in early 1979, has always regarded the situation in Indochina in a global context — as a test of its will against the Soviet Union's desire for global hegemony. In other words, its motivation for inflicting damage on Vietnam was to bleed the Soviet Union as well. Therefore, the lengthening of the bleeding process was to be welcomed and, in the absence of any alternative course to achieve the same end more effectively, there would be little reason for China to agree to a compromise.[14]

Finally, there is the persistent competition/conflict between Thailand and Vietnam. Thailand's post-war foreign policy seems to have revolved with fair consistency around the theme of containing Vietnam. Although Thailand is not necessarily a strong country in itself, it has shown a formidable ability to enlist international support to serve its own interests. It was Thailand that was instrumental in bringing into the region the United States, which really did have the strength to bleed Vietnam to death. When Vietnam took the gamble of invading Kampuchea, it was again Thailand that skilfully cooperated with ASEAN to muster the effective UN support that the five countries managed to gain, so causing the greatest humiliation that Vietnam has faced since its independence. What Thailand really wants is to neutralize Vietnam permanently, and as long as this is not possible, the present impasse may well be the next best choice. At least, it drains Vietnam's resources and stalls its economic development. That said, it seems likely that, barring some significant and unexpected change on the world scene, the present inflexible situation in Indochina will continue for some time.[15]

The Korean Peninsula

The cold war is still very much alive in the Korean peninsula, in form as well as in substance. Armed forces (some 1.35 million in all) are facing each other across the demarcation line of the demilitarized zone (DMZ), with an awesome arsenal of US nuclear weapons backing the South Korean forces. The array of fighter planes, tanks and artillery still has political relevance, in that it is widely accepted — in and out of the country — that the ultimate security of South Korea lies in military build-up. It is a situation that one would be more likely to associate with Europe than with Asia, where the danger of a large-scale armed showdown has been receding in the past decade or so.

The threat from the North, therefore, plays the most important role in the politics of South Korea, where the devastation of the war in 1950–3 is still vivid in memory. Successive governments have had to face the dilemma of how to encourage the process of popular political participation while maintaining a tight reign on internal security. The late President Pak Chung-hee was obviously caught between these contradictory goals. The spectacular growth of South Korea's economy during the 1970s, due in part to the success of his own economic policies, accelerated the clamour for a more open society and greater individual freedom. When, in 1972, he came down in favour of repressive measures, by way of the Yushin Constitution, it turned out to be a political disaster. A series of calamitous events followed, such as the abduction of Kim Dae-jung, the most outspoken opponent of the government, an assassination plot against the President and his wife, and recurrent student demonstrations which became increasingly threatening in tone. The sudden death of the President himself in October 1979 threw the country into a period of even greater confusion and instability. The pent-up frustration of the nation erupted in May 1980 in a massive uprising in the city of Kwangju, which had to be quelled by full-scale military intervention, leaving 189 people killed and many hundreds wounded. General Chun Doo-hwan, a leader of the martial law administration (the Standing Committee under the Special Committee for National Security Measures), subsequently emerged to take over the presidency, replacing the caretaker government of President Choi Kyu-hah, whose inability to govern had been proved by the uprising itself. Once again it had been made clear that, in South Korean politics, no leadership can effectively hold the country without the full backing and active involvement of the military.

Before taking the oath, General Chun staged a 'purification' campaign, purportedly to eliminate the 'corruption and irregularities of the old regime'. In the month of July alone, more than 8,000 allegedly corrupt civil servants were dismissed, while 1,800 officials of state corporations were removed from their posts. Even politicians of the highest rank were not exempt from the storm of 'purification'. Kim Dae-jung was arrested on the charge of sedition in connection with the Kwangju riots, and Kim Jong-pil, the former prime minister and the president of the ruling Democratic Republican Party, was forced to resign and to agree never to engage in politics again. The president of the opposition New Democratic Party, Kim Yong-sam, was put under house arrest, having also been forced to resign, and was asked to make a similar pledge to stay out of politics for ever.[16]

Having come to power in these circumstances, President Chun and his government knew that, if they were to generate credible support in the country, they had to offer a more positive and assertive set of policies than had been provided by the Pak regime, whose foreign policy was inflexible and somewhat isolationist as a result of its persistent anti-communist orientation. With an extremely efficient economic machine extending its tentacles throughout the globe, 40 million proud, energetic and highly educated people were crying out for a higher status in the world, a position commensurate with the growing power and wealth of the nation.

In this context, Chun's visit to Washington in January 1981 was an inspired move. He was received by the newly inaugurated Reagan administration, whose clear-cut anti-communism helped to straighten out the relations between the two countries. These had been somewhat strained by the emphasis on human rights of the Carter administration. In June 1981, the President made a two-week tour of ASEAN countries, the first ever by a Korean head of state. With an entourage which included a large group of businessmen, the visit made a fresh and favourable impression of a new Korea on the ASEAN countries, besides being useful in enhancing the President's own image back home. In August 1982, Chun travelled to Kenya, Nigeria, Gabon and Senegal, partly to expand South Korea's economic presence in Africa, and partly to undermine the position of the North, which had diplomatic relations with all these countries.

The decision of the International Olympic Committee to hold the 1988 Olympic games in Seoul is a timely development, with some far-reaching benefits for South Korea. First, it will give the nation a set of tangible goals to strive for over the next few years. Second, the entire process will give the government a golden opportunity to project a new image of the nation throughout the international community, an opportunity which the people of Korea have long been waiting for. Third, it will be a chance to launch a diplomatic offensive to attract as much international support as possible away from the North. And, last, like the 1964 Tokyo Olympics, which gave a considerable boost to Japan's economy, the multiplier effects of staging an international event of such scope and prestige in Seoul will be considerable.

The visit of the new Japanese Prime Minister, Nakasone Yasuhiro, in January 1983 was also well timed. Although the two-day working visit was too short for the two leaders to address themselves to the long-standing problems that existed between their countries, stemming from the 36 years of Japanese rule over the peninsula, Nakasone was able to straighten out some of the strains which he had inherited from his predecessor. Also, Japan's agreement to a $4 billion aid package, to be given over five years, was useful in reinforcing international confidence in the South Korean economy. Admittedly, South Korea's outstanding balance of external debt of $36 billion — not of the same order as those of some Latin American countries, but still sizeable enough — was not causing undue alarm in the Western banking community, mainly because of the extraordinary competitiveness which South Korea maintains in its export trade, but the Japanese pledge would nevertheless have given the Korean economic planners the elbow-room that they needed.

Although the political opposition was effectively removed by the 'purification' campaign, the root of the problem still remains. It lies not so much in the opposition itself as in the government's continuing difficulty in charting a course towards a more open society, while at the same time maintaining internal cohesion against the threat from the North. Speaking to the National Assembly on 18 January 1983, President Chun pledged that he would guide the nation into the ranks of the advanced countries before the end of his tenure — a goal which obviously entails developing genuinely democratic participation in the country's internal political process, as well as a wider base for its international relations.

As regards the latter aim, the government has been quite active in expanding its relations with the world, particularly with the communist countries — an effort which is often described as 'Nordpolitik', on analogy with West Germany's Ostpolitik. The hijacking of the CAAC airliner from China in May 1983 provided a welcome opportunity for South Korea to open a dialogue with China, which up till then had been consistent in its allegiance to North Korea. Another such opportunity arose in the shape of the Inter-Parliamentary Union Conference, which was held in Seoul in October 1983. Unfortunately, the visit of the parliamentarians from communist countries, from which so much had been expected, had to be cancelled at the last moment as a result of another airliner incident — the shooting down of a Korean Airlines plane by the Soviet air force — which took place shortly before the conference was due to start. The tragedy which befell the visiting President's entourage while on a visit to Burma, when a bomb-blast in Rangoon killed many of the finest men in the administration, was in a sense the result of Chun's positive diplomacy to court favours of the non-aligned nations.

* * *

The conclusion of the Burmese government's investigation of the Rangoon incident; which fully implicated North Korea in the killings, and its subsequent unilateral severance of relations provided a further link in the steady deterioration of North Korea's international image and standing. From having been a main battleground for the confrontation between communism and the free world and a birthplace of the cold war, North Korea — since the 1960s — has fallen into the chasm between China and the Soviet Union, and has once again become a 'hermit kingdom', though this time in a different context.

Following the hastily improvised North-South dialogue staged in the wake of the 1971 Sino-American rapprochement, North Korea launched a crash programme of economic development designed to move forward its Six-Year Economic Plan (1971-6) by two years. With the *Chuche* ('Self-reliance') philosophy, a kind of spiritual foundation to North Korean communism, imposed on all aspects of the people's lives, the situation was rather like a Great Leap Forward and Cultural Revolution put together. As in the case of China, the party hierarchy was extensively reshuffled, and during the 1970s many influential leaders, including Kim Yong-ju, the president's own brother, were dismissed. Entrusted to a group of young radical *Chuche* believers, the economy predictably fell into complete chaos.[17]

China is a cosmos in itself, and any irregularities in its behaviour will provoke instant comment. The turmoil in North Korea, by contrast, a small country with less than 16 million people and away from the focus of world attention, remained unnoticed throughout the decade, except for recurrent reports of North Korean diplomats who were apparently abusing diplomatic immunity to peddle duty-free liquor, jewels and drugs in various Western capitals — a reflection, no doubt, of the country's acute foreign-currency shortage, as well as the increasingly claustral behaviour of its government. One recent event, however, did cause some surprised international comment. This was when, in early 1983, President Kim Il-sung tried to secure for his eldest son, Kim Chong-il, automatic succession to power, reportedly against fairly solid opposition in the party and the government.

The Soviet Union has been feigning indifference to Kim's idea of hereditary succession to political power. China, however, has expressed strong reservations about such an unconventional idea for a communist country,[18] although it took great care not to push North Korea over to the Soviet side. China knew that its own latest policies to improve relations with Japan and other non-communist countries, as well as its all-out pursuit of economic development (tolerating covert dealings even with South Korea), were unsettling to North Korea, which felt more at ease with the orthodoxy of the Soviet Union. For one thing, unlike the Chinese, the Soviets have been consistent in supporting the North Korean demand for US withdrawal from the South. Also, given its halting relations with the Western world, North Korea would value economic and technological assistance from the Soviet Union. Furthermore, if only by reason of its geographical location, it is not in a position to oppose China openly, and must rely on Soviet protection. The visit of President Kim Il-sung to Beijing in September 1982 may have afforded an opportunity for some kind of pragmatic compromise between the two countries — a conjecture that is supported by the revelation, during Kim's stay, that Party Secretary Hu Yaobang had accompanied Vice-Chairman Deng Xiaoping to Pyongyang in April of that year to attend the celebration of the President's birthday. However, with its priorities firmly set on the Four Modernizations programme, it is highly unlikely that China would have encouraged whatever belligerent plans North Korea might have been harbouring against the South. As though by way of proof, during his visit to Japan in November 1983, Hu Yaobang made a special point of assuring the Japanese that the North Korean leaders had twice told him that they had no intention of invading the South.[19]

Hong Kong

Political uncertainty hangs heavy over Hong Kong. When the 99-year lease of the New Territories, as stipulated by the Second Convention of Peking in 1898, expires on 30 June 1997, it is generally assumed that China will reclaim sovereignty over Hong Kong Island and the Kowloon peninsula, which were ceded to Britain by two treaties, the Treaty of Nanking in 1842 and the First Convention of Peking in 1860. That the Beijing government is not in a position to compromise on the question of sovereignty is obvious. Hong Kong under foreign rule has long been a symbol of the insult and humiliation heaped upon China by the imperialists, and its recovery, which plays on the patriotism of every Chinese, communist and non-communist alike, has provided the Communist Party with an important means of rallying suport.

Beijing's attitude was made abundantly clear by its reaction to the statement made by the British prime minister, Margaret Thatcher, in September 1982, when she said that Britain still regarded the treaties as valid and binding. She added, perhaps in the after-glow of victory in the Falklands, that any nation which abrogated one treaty could not be trusted to keep others. It provoked an instant rebuff from the Chinese leaders, who retorted sharply that they would not recognize those treaties which were forced upon China by the imperialist powers.

However, in this case, unlike the Falklands dispute, Britain seems to have little desire for confrontation. It derives little economic benefit from governing Hong Kong, and the colony's future is not an issue in British domestic politics. In addition, the 1981 Nationality Act has denied to Hong Kong subjects any claim to residence in the United Kingdom, effectively exonerating Britain from responsibility for them. Likewise, Hong Kong students in British universities are classified as 'foreigners'.[20]

China, for its part, also wants a smooth transfer of sovereignty, since it hopes to retain Hong Kong's function as a regional centre of trade and finance. Deng Xiaoping himself has repeatedly stated that investors in Hong Kong should 'keep their hearts at ease,' and in January 1982 Prime Minister Zhao Ziyang told the visiting British junior foreign minister, Humphrey Atkins, that Hong Kong would continue to function as a free port and as an international financial centre.[21] In order to inject these statements with some legal credibility, a new constitution was adopted at the Fifth Session of the National People's Congress in November 1982, authorizing the Beijing government to establish special administrative regions (SARs). Although the name of Hong Kong was not specifically mentioned, the intention was clearly to enable Beijing to make any compromises that might prove to be necessary in the prospective negotiations with Britain on the future administration of Hong Kong.

Nevertheless, the notion of a capitalist Hong Kong operating under a communist administration is so startling that 5.15 million Hong Kong citizens find it hard to believe in it. Reflecting the pervasive sense of uncertainty, the Hong Kong dollar has never picked up since its dramatic fall in autumn 1982 and, in spite of booming export earnings in the first half of 1983, the government had to peg it to the US dollar in September, in order to fend off further downward pressure. Hundreds of people who want to emigrate flock every day to the consulates of the United States, Britain, Canada and Singapore. Nor have the negotiations now under way between Britain and China — talks began in July 1983, after Britain was reported to have made a conciliatory gesture on the question of sovereignty — done anything to allay people's fears.[22]

Understandably, Hong Kong citizens question whether the Chinese authorities are aware of the complexities of governing such a highly intricate socio-economic system as that of Hong Kong. Hong Kong has not become the place that it is thanks to a mere set of legal and administrative directives, but through the organic fusion of an exceptionally vital population, many of whom fled from the mainland, on the one hand, with the low-key, astute pragmatism of the British administration on the other. Whether such a delicate combination can be duplicated under communist rule is extremely questionable. Indeed, such attributes as freedom of buying and selling, unrestricted movement of people and goods, unhindered access to information through uncensored media, and many others which are vital to Hong Kong's functioning as it is constituted today, seem to go flatly against the basic premises of a communist state.

In an effort, perhaps, to restore people's confidence, on 15 November 1983, Ji Pengfei, Beijing's man in charge of Hong Kong and Macao affairs, tried to reassure a group of visitors from the New Territories not only by repeating the assertion that Hong Kong's existing system would remain unchanged, but by extending the duration of the proposed time-frame to fifty years after China's recovery of its sovereignty in 1997. He also said that the Beijing government had no intention of sending its own cadres into Hong Kong to meddle in the territory's affairs, and would leave the administration of Hong Kong entirely to its own people. He ended by saying that by 'existing system' he meant the basic arrangements of Hong Kong, including the freedoms of speech, publication and travel, and that such freedoms would be guaranteed by laws, which would serve as a 'mini-constitution' for Hong Kong after 1997.[23]

Although such pronouncements are welcomed by the people of Hong Kong, a sweeping statement of this kind tends to raise more questions than it answers. In the first place, for all Beijing's talk of a 'Hong Kong ruled by Hong Kong people', the people have never been allowed to participate in the negotiations between China and Britain that will determine their future. China has always been adamant about refusing the idea of popular elections in Hong Kong, at any level or of any kind. In particular, the notion of a plebiscite on future administrative arrangements has been anathema to Beijing. Apart from the communists' usual caution about any form of popular participation, this may reflect China's particular fear of the implications it could have for the question of Taiwan — a major outstanding issue in its programme. In any case, unless there is a drastic change in Beijing's attitude between now and 1997, the best that Hong Kong people can hope for, in terms of participation in the administration of their affairs, seems to be something like a Hong Kong-Chinese governor, approved, if not appointed, by Beijing, and a council, also screened by Beijing.[24]

Similarly, Ji Pengfei's talk about freedom is not very convincing. Freedom of travel in and out of Hong Kong could raise insurmountable problems of how to avoid a mass exodus from the mainland. Also, to have one region of the country, even if a special administrative zone, enjoy a privilege of overall freedom could cause unforeseen repercussions on the mainland's internal politics. In the event, would China be able to adhere to its commitment to Hong Kong and at the same time sustain its own communist system in the rest of the country? In this connection, an ominous note was struck by the Chinese communist press in Hong Kong, in which it was pointed out that while the wishes of the five million Hong Kong people would be taken into account, the will of the one billion people of China must prevail.[25] If, for whatever reason, Beijing should curb Hong Kong's freedom, even slightly, it would instantly destroy the business confidence which is the essence of its successful operation.

There is little, however, that either the British or the Hong Kong government can do about this. As long as Britain appears to have no intention of risking a major showdown with China in order to fulfil whatever moral obligations it may have in Hong Kong, the final decision on the management of the territory will have to be made by Beijing. Hong Kong residents and the foreign investors can only hope that the long-range economic and political interests of China, coupled with traditional Chinese pragmatism, will dictate policies to coincide with their wishes. In this connection, China's commitment to achieve a fourfold expansion of the economy within this century is a hopeful sign, since Hong Kong could play an invaluable role in helping China to meet this goal. Given the cultural and linguistic similarities, it can provide China with the system and know-how necessary to run a modern economy.

As though to prove the point, the recent expansion of China's Hong-Kong-related economic activities is significant. Following a Beijing directive of 1979, the Bank of China has been increasing its international business rapidly. Some 193 branches of the Bank and its associates throughout Hong Kong, Kowloon and the New Territories are now linked up by an on-line system, using IBM computers, which has increased its overall saving accounts dramatically, bringing its total overseas deposits to more than US $5 billion at the end of 1980 — a 39 per cent growth over the previous year. Also in 1979, it began to take an active part in Hong Kong's international loan market, its name appearing frequently in the syndicated loan advertisements with other multinational banks and investment houses.[26]

A similar thrust can be seen in trade and investment. More than fifty offices of China's state-run trading and investment corporations were reported to be operating in Hong Kong at the end of 1981, with an accumulated investment of somewhere between US $3 and $5 billion, while over 1,000 Hong Kong brokers and wholesalers and more than 100 department stores were said to specialize in the sale of mainland products. In the property market, China is reported to have invested between HK $3 and $10 billion in Hong Kong. In manufacturing, the Chinese have already set up a machine-tool and heavy-machinery plant, a shipyard, a cement plant and a watch-casing plant, some under joint venture contracts with US and Japanese interests. In 1978, a giant triangular deal worth some US $1.71 billion was concluded, whereby Britain provides power-generation plant, while China imports electricity for Guangdong province, in return for supplying coal to Hong Kong's China Light and Power Company.[27]

Table 6.1 shows the dramatic increase in China's net foreign exchange earnings from Hong Kong, which were estimated at US $6.9 billion in 1980, 36 per cent of its total foreign currency earnings. The highest increase occurred in the expenditure by visitors from Hong Kong, reflecting the increasing number of tourists who visit mainland China: from a total of 690,000 in 1976, the figure rose to one million in 1977 and leapt to 3.8 million in 1980. The entry 'Remittances and unrequited transfers' represents both money and gift parcels, and includes such consumer goods as TV sets, stereo equipment and various other home appliances sent from Hong Kong citizens to their relatives and friends in China. Also, many wealthy Hong Kong residents are known to have responded to requests from the mainland to make donations to schools, universities and hospitals.[28] Hong Kong's importance to China as a source of foreign currency earnings should not be exaggerated, because much of the trade surplus with Hong Kong comes from entrepot trade, and China would be able to continue such transactions with or without Hong Kong. However, as regards the remittances, tourist expenditures and investment profits, these are the direct result of

Hong Kong's own ability to create wealth, and one may reasonably assume that China would think twice before giving up such a lucrative source of income.

Table 6.1: *China's Net Foreign Exchange Earnings from Hong Kong (US$ million)*

	1977	1978	1979	1980
Trade surplus	1,741.4	2,259.5	3,034.1	4,407.2
Remittances and other unrequited transfers	394.9	477.3	562.9	673.7
Expenditure by HK visitors	223.5	367.5	819.4	951.8
Investment profits	367.8	461.0	609.9	824.9
Total	2,727.6	3,565.3	5,026.3	6,857.6

Source: *Far Eastern Economic Review*, 20 January 1983, p. 42.

Hong Kong's investment in China has also increased since 1979, when the new code on joint ventures was enacted, and greater autonomy granted to Guangdong and Fujian provinces on policies concerning foreign trade and investment. The designation of four cities in the southern Chinese provinces, namely Shenzhen, Zhuhai, Longhu and Huli, as special economic zones (SEZs), served to accelerate this trend. Between 1979 and the end of 1982, Shenzhen alone had a total of HK $9.2 billion pledged for investment for 1,500 projects.[29] Property investment was estimated at about HK $1.04 billion as of May 1981, and the purchase of residential property is spreading out of the SEZs to other major cities such as Shanghai, Hangzhou and Fuzhou — enterprises in which a number of Taiwanese are said to have participated. Joint ventures in the hotel industry seem to be especially attractive to Hong Kong investors, with amounts reportedly reaching about US $170 million as of mid-1980.[30]

Industrial output in such fields as electronics and textiles in the southern Chinese provinces of Fujian, Guangdong and Guanxi, which have absorbed the bulk of overseas investment, is expected to grow substantially during the 1980s. Fujian is busy improving its infrastructure in energy and transportation, with a view to substantial industrial growth after 1985. (The proximity of Fujian to Taiwan, and the historical ties between the two, are significant, and the Chinese leaders seem to be conscious of the effect which successful economic development in Fujian might have on the issue of reunification with Taiwan.) Guangdong is concentrating on growth in such industries as power generation, petroleum, chemicals, building materials and non-ferrous metal extraction. The construction of a motorway connecting Hong Kong, Shenzhen, Guangdong, Zhuhai and Macao is reported to be starting in autumn 1984.[31] With this extensive industrial expansion in southern China in view, a number of foreign banks, such as the Hong Kong and Shanghai Banking Corporation and Takugin International (Asia) Ltd, have already opened branches in Shenzhen, while many others are waiting for approval from the Bank of China.[32]

It is interesting to speculate on why China has chosen to carry out such a thoroughgoing experiment in the ways of free-market economy at this juncture. One obvious reason is to prepare the mainland economy for the prospective integration of Hong Kong, so as to ease the shock and enable it to maximize its gain from integration. Obviously, though, an exercise of this kind involves very great risks for a communist government. The germ of the market economy, with its seductive paraphernalia of private ownership, individual freedom and the like, could well contaminate the hearts and minds not only of the masses but also of the cadres and officials of the Chinese system. Like the tail wagging the dog, a tiny Hong Kong could precipitate a groundswell of change and shake the very foundations of communism in China. At the least, it involves a danger of alienating the more ideologically orientated Chinese cadres, who might see this experiment as a betrayal of the cause as well as a threat to themselves.

* * *

The question of Hong Kong is not unconnected with that of Taiwan, as yet another newly industrializing country (NIC) which suffers from an anomaly in global politics. In fact, the long-range Chinese goal of reunification with Taiwan may work to the short-term benefit of Hong Kong, because the consideration of Taiwan could give China a strong motive to moderate its policy towards Hong Kong. If China blunders in Hong Kong and recreates the trauma of rout and exodus that took place after the communist victory in 1950, the peaceful reunification of Taiwan would be put off for many decades.

In Taiwan's case, sovereignty is not the issue. The Kuomintang (KMT) government of the Republic of China (Taiwan) is in complete agreement with the communist government in Beijing on China's sovereignty over Taiwan. Rather, the KMT contends that it is the legitimate government of the whole of China, temporarily prevented from exercizing its power by the unlawful occupation of the mainland provinces by the communist rebels. True to this contention, three decades after its removal to Taiwan, the KMT still maintains the form and appearance of the government of all China, with more than two-thirds of the seats in its parliament, the Legislative Yuan, being occupied by mainlanders who were elected to these posts before 1950, and who were 'frozen' in office pending the KMT's eventual return to the mainland.

Ever since President Nixon's overture to Beijing in 1971, Taiwan has experienced a steady erosion of its international status, with many countries shifting their allegiance to Beijing. In order to cope with a difficult situation, as well as to preserve its all-important trade relations, the KMT government swallowed its pride and approached many of these countries with proposals for retaining economic relations. In view of Taiwan's successful economic performance, these proposals were generally welcomed, and it currently maintains active economic relations with 150 countries around the world, even though it has diplomatic relations with only 22 countries.

Like other East Asian NICs, Taiwan recorded a phenomenal rate of economic growth (average 10 per cent annually) during the 1970s. Its per capita income of US $2,360 in 1981 was the fifth largest in Asia, following Japan, Hong Kong, Singapore and Brunei.[33] Moreover, unlike many other fast-developing countries, its government was successful in minimizing the gap between rich and poor, a major achievement of KMT economic policies. This and other successes in domestic policy reinforced the KMT's popularity in Taiwan, even though it is essentially a government of mainlanders ruling over the indigenous Taiwanese population with the aid of martial law. In the election for the Legislative Yuan held on 3 December 1983, the KMT won more than 70 per cent of the total ballot, while, in spite of the relative freedom granted for overall political activities, the opposition Tangwai ('outside the party') groups lost two of the seats they had won in 1980.[34]

The ultimate security of Taiwan lies with the United States, although it seems highly unlikely that, in the foreseeable future, China would attempt to take it over by force. Also, one can perhaps argue that, since the United States formally recognized Beijing as the sole legal government of China, as of 1 January 1979, the question of Taiwan's security has changed its basic character from that of a frontline issue in the East-West confrontation to a problem that is essentially internal to China. In October 1981, Marshal Ye Jianying made a nine-point peace proposal, under which Taiwan would be allowed to retain its present social, economic and political system (even to keep its own armed forces) after reunification. As was to be expected, President Chiang Ching-kuo of the Taiwanese government rejected the proposal outright as a typical communist ruse. He commented later that, while his government, too, was eager to see China reunified, it could be done only under the banner of Dr Sun Yat-sen's Three Principles of the People.[35]

In view of the differences in their political systems, economic performance and living standards, it will be very difficult to find a means of reunifying these two regimes. However, given the uncanny pragmatism of the Chinese people, the strikingly similar attitudes and behaviour demonstrated by their leaderships and, above all, the extraordinary pace and scope of Beijing's experimentation with market-oriented economic policies, perhaps one should not rule out the possibility entirely. In any event, the way China solves the problems of its peripheral entities — a legacy of its colonial and ideological past — will have important implications not only for the political and economic evolution of the region, but also for the future of the global ideological confrontation.

Communism in Asia

The Third Indochinese War demonstrated what difficult problems communism in Asia faces. The phenomenon which can be described as a chain reaction of hegemony — a procession of events whereby China reacted against Soviet hegemony, and Vietnam accused China of hegemony while it was itself adopting a hegemonic attitude towards Kampuchea — revealed the limits of communism as a political proposition. Although the circumstances, as well as the manner, of hegemonic behaviour in each case were different, the overall implication was that in the communist world equal and pluralistic relationships are not possible. This revelation was a considerable blow to communist movements and leftist thinking throughout Asia.

In Japan, for example, left-wing idealism increasingly lost its powers of persuasion, so much so that it became one of the causes of the growing conservatism of Japanese politics — as shown by the LDP's increased majority in the 1980 double elections. Even in the December 1983 election, when the conservatives lost heavily because of the political morality issue, the JCP lost some of its seats, and even the JSP did not do as well as other middle-of-the-road parties in wresting votes from the LDP. In the Philippines, to take another example, the New People's Army seems to be having difficulty in gaining momentum in spite of social and economic problems that would normally be conducive to revolution. In Thailand, too, many of the young radicals who joined the underground communist movement in the jungles are surrendering and being reintegrated into society. The Thai Communist Party organization is split by internal feuds and no longer possesses the vision or blueprint for the future that once inspired the country's youth.[36]

In the 1920s and 1930s, communism provided ideals and aspirations for many Asian revolutionaries. The neatness of the Marxian concept of the world, which states that the profit and power of one group of people are paid for by the loss and misery of other groups, struck a chord in the minds of those who were longing for political liberation and an end to misery, oppression and exploitation. Furthermore, communism afforded them a self-contained political proposition. To the individual, it gave the valuable psychological fortification of being part of a global struggle for the liberation of man, rather than being a lonely rebel against the overwhelming power of government, whether colonial or otherwise. To the group, it offered a technique of revolution, ranging from a method of organizing the masses into a strategic front and conducting effective propaganda, through to the tactics of selective terrorism.[37]

However, the context of the anti-colonialist and independence movements in Asian countries was fundamentally different from that of Marx's original 'proletarian revolution'. These Asian movements had a strongly nationalistic orientation, and as such, in order to be effective, had to involve the national bourgeoisie, regardless of its social origins. This was clearly contrary to the concept of class war and incompatible with international proletarianism. During the 1920s there was debate within the Soviet leadership about whether to give full-scale support to these independence movements. On the basis that the first priority was to weaken the capitalist European countries and the United States, it was decided that they should be supported — in the belief, in the words of Lenin's apocryphal statement, that the road to Paris lay through Peking. In retrospect it seems that, since the nineteenth-century industrial relationships which had bred Marxism had already disappeared in Western Europe and an orthodox proletarian revolution was difficult to bring about there (witness the revolutionary failures of 1917–19), the Soviet leadership was forced to consider the 'indirect' undermining of Western Europe through the backdoor of its colonies.[38]

In this way the revolutionary movements in China and other Asian countries, nurtured and supported by the Soviet Union and its revolutionary arm, the Comintern, gave a decisive impetus to the anti-colonialist independence movements. The convulsions of World War II and the anti-Japanese resistance movements throughout the continent were particularly opportune for generating further momentum for anti-colonialism, and as a result, in the post-war period, as the Soviet Union had desired, the domination and exploitation of the resources of the Asian countries by the European powers came swiftly to an end.

Further, in the 1950s and 1960s, these independence movements succeeded in involving the United States in a series of prolonged local wars which dealt a blow to its military, social and economic structure that surpassed the Soviet Union's wildest dreams. This was a timely outcome for the Soviet Union, which had suffered from a costly military impasse in Europe and was rapidly losing its appeal and hence its ability to lead the world communist movement. The continuation of the 'revolutionary' wars in Asia performed the double function of helping to sustain the image of communism as a revolutionary movement long after it had ceased to be one, and of upholding the prestige of the Soviet Union as one, at least, of the centres of revolution.[39]

However, the great military machine of the Soviet Union, such an effective means of control in Eastern Europe, seems to be ineffective when it comes to controlling the Asian communist countries. Thus, there was nothing the Soviet Union could do to stop the fierce and scornful criticism which China heaped upon it from the 1960s onwards, charging it with 'great-power chauvinism' and 'revisionism'.[40] After the Sino-American rapprochement, when China openly identified the Soviet Union as its principal enemy and attempted to form a world-wide anti-Soviet alliance, the Soviets could not lift a finger in opposition. Only in Vietnam, where there was a strong fear of Chinese hegemony, could it maintain some influence through its unusually generous assistance; and even then, in deference to Vietnam's strong nationalism, the nature of this influence was different from the pervasive domination exerted over the East European countries.

With the exception of Japan, where the concept of proletarian class struggle has had some relevance since the 1920s and 1930s, all the communist movements in Asia developed in the context of the struggle for independence. As a result, even when these regimes come to power, they retain a strong orientation towards nationalism as well as indigenous cultural traditions. Pham Nhu Chuon, the vice-chairman of the Vietnamese State Committee of Social Science, stated in Bangkok in June 1980: 'To the Vietnamese people, the truth that "nothing is more precious than independence and freedom" has become a historic experiment, a part and parcel of our flesh and blood, and the value of all values.'[41] If this statement represents the feelings of most Vietnamese, then what it means is that Vietnam is prepared to fight anybody, be it China, the Soviet Union or the United States, in order to protect its independence.

In the case of Vietnam, communism was at least useful in giving a conceptual framework to its struggle for independence. However, the introduction of communism will have varied results, depending on the political and cultural situation of the country, and sometimes unexpected ones. One such example is the excesses of the Pol Pot regime, which almost annihilated the people and culture of Kampuchea. It has been argued that the barbarities of Pol Pot's Khmer Rouge led to a phenomenon of atavism, whereby an exposure to communism and its ruthless power system had the effect of awakening certain aspects of the ancient and long-dormant Hindu-Khmer political culture.[42] In any event, it would be exceedingly difficult to bring together these countries of diverse cultural traditions in order to develop some sort of 'communist international'.

The multifarious development of communism in Asia is likely to make the basic political structure of the region increasingly different from that of Europe, where 'the two alliances confronted each other with the greatest concentration of military power anywhere in the world,' and where, 'contrary to the hopes of the 1970s, the confrontation beween the Soviet Union and the West is going to stay both in terms of a power struggle between two superpowers present on European soil and in terms of a conflict between two irreconcilable ideologies, that of a totalitarian system and Western democracy'.[43]

Even in Europe, however, the cold war as a purely ideological force seems to be losing momentum. The Soviet Union, with its huge bureaucratic organization and military establishment, would perhaps see the relevance of the cold war more in terms of protecting the state apparatus, and perpetuating the vested interests of the party cadres, the government and the military officials, than in terms of spreading the ideology of communism to the ends of the earth. Viewed in this context, the Soviet threat to Western Europe could well be self-perpetuating. The fact that the Western countries, whatever the temporary problems of depression, are enjoying freedom, prosperity and a functioning democracy is in itself a challenge for, and a subversive influence on, the satellite countries of Eastern Europe and ultimately the Soviet Union itself. Therefore, as long as there is Western Europe, the Soviet leadership cannot afford to lower its guard. It must use the methods of coercion, intimidation and persuasion, and, as far as possible, try to promote the 'Finlandization' of Western Europe.[44]

In East Asia, however, with the possible exception of sparsely populated Mongolia, the Soviet Union has no satellites, and therefore does not need to be so alert to the danger of subversion from outside. As a result, it can afford to be more relaxed than it can in Europe. As for China, whether because it has no satellites, or because it has complete confidence in its own culture, it shows no signs of defensiveness towards the market-oriented economies of such countries as Hong Kong, South Korea and Japan. On the contrary, it seems to be eager to learn from their success.

Admittedly, the Soviet Union's military presence, which literally encircles the globe, does pose a threat to the Asian Pacific region. Bases such as Cam Ranh Bay strengthen its navy's route from the Indian Ocean to the Japan Sea; the increased activities of its carriers and submarines in the Western Pacific constitute a threat to the waterways and sea-lanes of the whole region. Also, the latest talk of the deployment of SS-20s in Asian Siberia is frightening, at least psychologically. On the other hand, it is perhaps safe to assume that, if the Soviet Union were to consider a showdown with the West, which might very well lead to World War III, it would have to do a hard calculation on the possible reaction and response of the communist nations in Asia. With ideological allegiance rapidly losing its relevance, the Asian communists may not feel obliged to offer blind cooperation. A proposition to assist the Soviet Union to establish its hegemony over the rest of world would hardly appeal to the Chinese, for example. It is even possible that the large Soviet forces assembled on the Sino-Soviet border and in eastern Siberia are gathered there with a view not so much to attacking Asia as to neutralizing it in the event of a global war. It would be ironical indeed if communism in Asia, which was nurtured by the USSR in order to undermine the power base of Western capitalism, turned out to be an Achilles' heel for the Soviet Union if it should decide on a final confrontation with the West.

Chapter Seven

THE REGIONAL ECONOMY AND INTERDEPENDENCE WITH JAPAN

In the past decade, the Asian Pacific region has greatly increased its weight and influence in the world, if only because of its remarkable economic growth. Following Japan's astonishing record of growth in the 1960s, the NICs of the region, namely South Korea, Taiwan, Hong Kong and Singapore, displayed an equally remarkable rate of growth in the 1970s. They were closely followed by the four less-developed members of ASEAN, Indonesia, Malaysia, the Philippines and Thailand, which achieved annual growth rates of between 6 per cent and 8 per cent consistently throughout the 1970s. As a result, the region currently accounts for one-sixth of total world trade, and recently overtook Western Europe as the biggest trade partner of the United States. It was the first time in modern history that a group of non-Western countries had achieved such a substantial economic success, and it is the first time that a group of developing countries has managed to demonstrate such a sustained pattern of growth (see Table 7.1).

The population of the region, which exceeds 1,500 million, over one-third of the world total, consists largely of Mongolian and Malay stocks, with complex ethnic variations. The region covers a major part of continental East Asia, plus a vast number of islands and peninsulas which are loosely connected by a great network of ocean- and sea-lanes. In contrast with Western Europe or Latin America, there are no cultural traditions or historical experiences to bind the countries of the region together, and with the exception of ASEAN, there are few forums for regional dialogue or consultation. Rather, each country tends to go it alone, with little regard for the policies of the others. This lack of coordination has resulted in conflicting policies and measures to the detriment of economic performance.[1]

For all this diversity, however, the non-communist countries of the region — from Japan to the NICs to ASEAN — are committed to the goal of market-oriented economic development, and have pursued it energetically, with forward-looking and, by and large, rational policies. There is a core set of similarities in the way industrialization is taking place among these countries. Their bureaucracies are in general efficient and ably staffed, and there is a sense of partnership and shared goals between business and government in most countries — not only in a racially homogeneous country like Japan but also in socially fragmented nations like Malaysia and Indonesia.

Table 7.1: Demographic and Economic Statistics for Selected Asian Countries

	Population (1981 millions)	Area (1,000 km^2)	GNP (1981)		Annual growth rate 1977-81 (%)
			(US$ billion)	Per capita (US$)	
South Korea	39.33	98	66.8	1,720	6.5
Taiwan	18.30	36	45.0	3,587	8.6
Hong Kong	5.23	1	28.1	5,390	10.9
(Sub-total)	(62.86)	(135)	(139.9)		
Indonesia	154.66	2,027	77.7	520	8.2*
Philippines	50.74	300	39.7	790	6.2*
Thailand	48.49	514	36.9	770	7.3
Malaysia	14.14	330	25.8	1,820	7.8*
Singapore	2.47	0.6	12.8	5,220	8.7†
(ASEAN total)	(270.50)	(3,172)	(192.9)		
Japan	118.45	378	1,139.3	9,684	4.9
China	1,020.67	9,597	328.0	321	—

*1977-80   †1976-80
Source: Keizai Koho Center, *Japan 1983* (Tokyo, 1983), pp. 1 and 9.

Perhaps the spirit of striving stems from the sense of vulnerability and urgency that seems to prevail in the region. For divided nations like South Korea and Taiwan, economic success is imperative if they are to maintain their status and influence in the international community against the consistent pressures and threats of their communist counterparts. (One is put in mind of the pervasive sense of insecurity and urgency which haunted the Japanese in the Meiji era, and drove them into an inexorable pursuit of economic as well as military power.) Likewise, Singapore tends to regard economic success as a prerequisite for its survival as a small and predominantly Chinese city-state floating in the sea of the Malays. For the governments of the less-developed Southeast Asian countries, too, economic success has long been considered a *sine qua non* of social stability and resilience — essential conditions for combating the recurrent threat of internal dissidence and insurgency.

By 1983, most of the Asian Pacific economies had recovered from the effects of the prolonged world recession which followed the 1979 oil shock. Helped by the upturn in the US economy, the East Asian NICs recorded spectacular rates of growth in their exports in the first three quarters of 1983 (over the same period in the preceding year): 34 per cent for Hong Kong, 33 per cent for South Korea, 30 per cent for Singapore and 27 per cent for Taiwan. Domestic demand also rose, reflecting the low inflation rate prevalent in the region, 1983 figures being 2 per cent for Singapore, 2.3 per cent for Japan, 4.6 per cent for Taiwan and 4.7 per cent for South Korea.[2] The less-developed countries of ASEAN have generally benefited from the rising prices in such commodities as rubber and palm oil, although Indonesia is still having difficulties as a result of the downturn in oil prices. Malaysia recorded a 22 per cent increase in manufacturing exports, mainly electronic components, showing a somewhat similar pattern of recovery to that of the NICs, while Thailand expected a robust 6.5 per cent growth in manufactures for 1983 on account of the booming domestic demand in such items as cars, motorcycles, home appliances and processed foods. Although political uncertainties shadow the Philippines and Hong Kong, the region as a whole seems to be in for another spell of growth.

No doubt further steps will be made in terms of both interdependence and expansion during the remaining decade and a half of this century, though presumably not at such a rapid pace as in the preceding decade. Although a period of adjustment lies ahead, most of the countries, both developed and developing, are in a strong position to cope. Having sustained a considerable economic momentum and having already made substantial structural adjustments, they are likely to emerge from the current recession with leaner and more competitive export industries. More efficient use of economic resources, notably as regards investment costs and energy supplies, will be a central feature of economic policy in all countries, and there will continue to be an emphasis on labour-intensive manufacturing exports in the developing countries. The latter could help reduce investment costs, as well as easing balance-of-payments problems and increasing employment. Although domestic policies will be the key determinant of how well the Asian Pacific countries

The Regional Economy and Interdependence with Japan 139

perform individually, there will no doubt be growing pressure for greater Japanese economic support in the region.

The rest of this chapter deals, first, with the economies of ASEAN as relatively successful models of economic development; second, with the concept of Pacific Economic Cooperation, which is attracting increasing attention both inside and outside the region; and, finally, with the way in which Japan involves itself in the regional development process, with special emphasis on its activities of trade, investment and aid.

ASEAN

The population of ASEAN, totalling 270 million in 1981, roughly equals the 271 million of the European Community (although its land area of 3,172 km^2 almost doubles the EC's 1,660 km^2). In contrast with the EC, however, it is characterized by the diversity — in the composition, size and structure — of its members' economies. The smallest country of the five, Singapore, has by far the most advanced economy, with a per capita GNP of $5,220 in 1981, whereas that of the largest, Indonesia, hovered around a meagre $520, less than one-tenth of Singapore's. It is as though there were a negative correlation between the size of the population and the level of economic development where ASEAN countries are concerned (see Table 7.1). In Singapore, where high-wage/high-productivity policies — designed to drive out labour-intensive industries — have been applied, the manufacturing industry is rapidly moving into more knowledge- and technology- intensive sectors.[3]

Endowed with a vast land area and rich natural resources, Indonesia certainly has the potential for development. However, it is still beset by a host of problems, which range from a lack of adequate infrastructure, such as electricity and transportation, to the uneven distribution of the population, which results in a frightening density in the islands of Java and Bali, while the vast and resource-rich outer islands are only sparsely populated. Since the 1973 oil crisis, its economy has been boosted by the massive production of oil and gas, and hence saved from the problems of balance-of-payments deficits. However, the latest fall in energy prices is causing considerable difficulties for Asia's only OPEC member. The rupiah was devalued by 28 per cent at the end of March 1983, and many ambitious industrial projects have had to be cancelled or postponed. The growth rate in

1982 dropped to 3 per cent, and it does not look as if 1983 has done much better. Part of the difficulty stems from the delay in an attempt to diversify the economy. In consequence of vigorous government measures, manufacturing is growing rapidly, though its share of GDP was still only 9 per cent in 1979, the smallest figure in ASEAN (see Table 7.2).

Table 7.2: *Structure of the Production of ASEAN Countries, 1979 (% share of GDP and annual growth rate)*

	Agriculture		Industry				Services	
			Total Industry		Manufacturing			
	Share	Annual growth 1970-9	Share	Annual growth 1970-9	Share	Annual growth 1970-9	Share	Annual growth 1970-9
Indonesia	30	3.6	33	11.3	9	12.5	37	9.2
Malaysia	24	5.0	33	9.9	16	12.4	43	8.4
Philippines	24	4.9	35	8.4	24	6.7	41	5.4
Singapore	2	1.7	36	8.6	28	9.3	62	8.5
Thailand	26	5.4	28	10.4	19	11.4	46	7.7

Source: Lawrence B. Krause, *US Economic Policy Toward ASEAN* (The Brookings Institution, Washington, 1982), p. 18.

Malaysia, the second smallest member both in land area and in population, recorded a per capita GNP of $1,820 in 1981. It, too, has a rapidly growing manufacturing industry, although, being the world's largest producer of natural rubber and palm oil, it earns roughly a quarter of its GDP from agriculture, which, with the shrinking demand and falling prices in the world market, poses problems for the government.

The Philippines and Thailand, both of which have a population of about 50 million, approximate also in the size of their economies: GNP in 1981 was estimated at $39.7 billion and $36.9 billion, respectively. The Philippines' share of manufacturing in its GDP was 24 per cent in 1979, second only to Singapore, with exports of items in the electronics and garments industries increasing at an impressive rate each year. The country's traditional agricultural exports, however, such as sugar and coconut, which are still dominant, have been hit by the world recession and are causing considerable balance-of-payments difficulties. Foreign debt reached over $16 billion at the end of 1982, and in October 1983, as the debt mounted further, a temporary moratorium on debt repayments had to be instituted. Politically, the Philippines is the most troubled of the ASEAN nations. The opposition leader, Senator Benignò Aquino, was assassinated in August 1983; the reported illness of President Marcos is adding uncertainty to the succession problem; and the forces of rebellion, ranging from the dissident Catholic clergy to violent Muslim and communist insurgents, are gaining ground and undermining the confidence of foreign creditors. The Philippines' share of intra-ASEAN trade has been traditionally small, reflecting, perhaps, a geographical location slightly detached from the rest of ASEAN, as well as a historical dependence on trade with the United States and other extra-regional markets.

As for Thailand, in spite of continued fighting in neighbouring Kampuchea and frequent changes of government, the economy showed remarkable resilience in 1982 and 1983, with the prospect of smaller trade deficits and a considerably lower inflation rate. This was partly due to the bumper harvest of rice in 1981, which led to the export of very nearly 4 million tonnes in 1982, making Thailand the largest rice exporter in the world. The diversified range of its export products, embracing industrial raw materials, manufactures and foodstuffs, helped Thailand to pass through the world recession relatively unscathed, in contrast with Malaysia and Indonesia, which depend heavily on such standard commodities as tin, rubber, palm oil and petroleum products. Imports are expected to drop, partly owing to the depression in domestic economic activities, which also had the effect of pushing down inflation to the projected rate of 5 per cent. A slow but steady increase in the production of natural gas in the Gulf of Thailand — made possible by the prudence of the government, which resisted the temptation to spend the money before it came in — makes Thailand's economy one of the most promising in the region.[4]

The geographical distribution of ASEAN trade is shown in Table 7.3. The largest trade partner is Japan, with a turnover of $32,369 million, or 23.9 per cent of total ASEAN trade, in 1980. Next comes intra-ASEAN trade, with 17.1 per cent, and then the United States and the EC, whose shares were 15.9 per cent and 12.5 per cent respectively.[5] With the exception of Japan, which still imports oil and gas from the region in quite substantial quantities, the shares of OECD countries have been on a slight but steady decline since the 1970s — presumably a reflection of ASEAN's efforts to diversify its markets and supply sources. The rise in intra-ASEAN trade which has resulted from this trend is still short of expectations, owing, perhaps, to the basically competitive, rather than complementary, nature of the ASEAN economies.

Table 7.3: ASEAN Trade Partners (Exports plus Imports) (US $ million)

	1974	1976	1978	1980
Intra-ASEAN	6,528 (14.1)	8,199 (15.4)	11,523 (15.5)	23,182 (17.1)
OECD	29,325 (63.2)	32,769 (61.5)	44,983 (60.5)	78,120 (57.8)
EEC	6,498 (14.0)	7,649 (14.4)	10,507 (14.1)	16,842 (12.5)
Japan	12,366 (26.7)	12,736 (23.9)	17,701 (23.8)	32,369 (23.9)
US	7,799 (16.8)	9,499 (17.8)	12,719 (17.1)	21,455 (15.9)
OPEC	3,624 (7.8)	4,401 (8.3)	6,265 (8.4)	12,891 (9.5)
Socialist bloc	704 (1.5)	612 (1.1)	839 (1.1)	1,635 (4.6)
Others	6,186 (13.3)	7,304 (13.7)	10,787 (14.5)	19,407 (14.4)
Total	46,367	53,285	74,397	135,235

Note: Figures in parentheses are percentages of total.
Source: Institute of Southeast Asian Studies, *Southeast Asian Affairs 1982* (Singapore, 1982), p. 35.

In view of the good past performance of the group as a whole, international institutions are making rosy predictions for the 1980s. The Wharton School of Economics forecasts that the ASEAN countries will continue to show a combined rate of growth which surpasses that of the East Asian NICs until 1985, and that its inflation rate will go down considerably. Chase Econometrics views Malaysia as the fastest grower in the region, but is somewhat pessimistic about the Philippines and Thailand.[6] If the global economy continues to recover, ASEAN is likely to be one of those that can make the most of the improved situation; if recovery is short-lived, it will probably be able to minimize the damage, through active interaction and cooperation with other lively economies in Northeast Asia, which could well include China.

On 1 January 1984, Brunei became fully independent, severing its last formal links with Britain, which since the 1959 Constitution had been responsible only for defence and foreign relations. Brunei, sandwiched between the East Malaysian states of Sabah and Sarawak, is well endowed with large oil and gas reserves, accounting for 98 per cent of its export earnings and 78 per cent of its GDP.[7] The country's oil production is managed jointly by the Brunei government and Shell, the giant energy company, while in the case of gas the Japanese are heavily involved, taking all Brunei's LNG exports. The extent of Japan's interest was demonstrated by Prime Minister Nakasone's visit to the state in May 1983.

Brunei refused to join Malaysia in 1963, and only at the end of the 1970s did a thaw in relations between the two countries take place; now, however, problems with both Malaysia and Indonesia seem to have been ironed out. There was no serious opposition within ASEAN to expanding the organization to include Brunei, which has had observer status since 1981 and had expressed a desire to join on gaining full independence. (And, indeed, on 7 January 1984, only a week after gaining independence, it became a formal member.) Economically, Brunei attracts the active attention of ASEAN member countries, as shown by the visit of a high-level delegation from Thailand in November 1983, led by Foreign Minister Siddhi Savetsila. Thai business and financial circles are keenly interested in Brunei's extraordinary wealth and its capacity for earning foreign currency, and hope to attract some of this business to Bangkok.

* * *

A crucial challenge to ASEAN countries in the coming years will be the political and socio-economic problems which follow a rapid process of development. Some people argue that a developing country inevitably moves into a 'zone of instability' when its per capita income reaches a level between $300 and $1,000. Although Singapore has long passed this stage and Malaysia is graduating from it, the three large ASEAN economies are right in the middle of it. The problems which are usually associated with such a 'zone' include the unequal distribution of wealth, unemployment, urban problems, a growing number of labour disputes, nationalist reaction against foreign investment, conflicts between various political and other interests, as well as difficulties in making managerial ability keep pace with development.[8]

These problems exist in many of the ASEAN economies. They are closely connected with traditional socio-political problems, such as those arising, for example, from the form of asset ownership. In some countries, the bulk of assets is owned by a handful of families, which leads to unequal employment opportunities. In others, where economic success has led to asset ownership on an international scale, the problem of unequal income distribution becomes more acute, because, in the absence of progressive taxation, the fruits of development tend to benefit only the wealthy classes. Also, except for Singapore, the population of all the other ASEAN countries has increased more than 2 per cent annually during the 1970s, which will perpetuate the problems of rural-urban migration and urban unemployment.[9]

Although the private sector has played a key role in the development of ASEAN economies, the governments are sufficiently centralized to exercise decisive control over every phase of economic management. Therefore, any shortfall in the social objectives of development can reasonably be blamed on insufficient political will or lack of effective management on the part of governments. With people's expectations growing, any signs of failure to achieve these objectives can create political problems, as seen in the recent difficulties in the Philippines. Many ASEAN governments are so concerned about such issues that they are giving the elimination of poverty overriding priority in their current economic programmes, as shown by Malaysia's New Economic Policy and Indonesia's Third Five-Year Plan (Repelita III). Coupled with other difficult problems, such as corruption in high places, succession issues, potential conflicts among different ethnic and religious groups, which all the members of ASEAN have to face at one time or another, these matters of social adjustment will constitute the main challenge to ASEAN as it moves towards its avowed goal of development and modernization.

Pacific Economic Cooperation

The rapid growth of the Asian Pacific economy during the 1970s has led to a proposal for cooperation in a wider area — namely, the Pacific Basin as a whole. This would include the five Pacific OECD countries (Canada, the United States, Japan, Australia and New Zealand), three East Asian NICs (South Korea, Taiwan and Hong Kong) and the five ASEAN countries (Indonesia, Malaysia, the Philippines, Singapore and Thailand), as well as a group of Pacific island states. The area's consistently robust growth seems to distinguish it from many other parts of the world, where slow growth, increasing unemployment and a foreign debt of vast proportions persistently hamper development.

The fact that the countries in the Pacific Basin are at different stages of development tends to increase interdependence, as the expansion of intra-regional trade over the past decade demonstrates. Intra-Pacific trade now accounts for more than half the total trade of Hong Kong, Indonesia, Malaysia, Singapore and Thailand, and for approximately one-third of that of the Philippines and Indonesia. The dependence on regional outlets for goods and resources has expanded quite rapidly — from very low bases in some countries. In Hong Kong, for example, where only 14 per cent of total exports went to East Asian developing countries in 1972, the figure had risen to 21 per cent by 1981. Or, to take a specific example, in the early part of the 1970s most electronic component exports in the region went to the United States and Europe; by the mid-1970s, the picture had greatly changed, with intra-regional trade in semiconductors, integrated circuits and finished products having increased sharply. By 1974, Japan was sending more than one-third of its integrated circuits to other Asian countries, and by the end of the decade well over two-thirds. By 1980, Japanese-South Korean bilateral trade in electronics had reached $500m.[10]

This trend has contributed to the evolution in the region of an intra-industry, or vertical, division of labour in the electronics and consumer electronics industries, whereby — simply put — goods with the highest value-added are produced in Japan, while other Asian countries concentrate on various related, but less sophisticated, aspects of production, for which cheaper labour, raw materials and energy are more important than knowledge-intensive technology. While this phenomenon is most prominent in the electronics industry, it is also seen in textiles (as demonstrated by the spread of Japanese multinational firms such as Teijin and Toray).[11] The same thing has happened to a certain extent in non-ferrous metal production, where Japanese investment promotes the development of aluminium or steel facilities abroad and then reimports the goods for manufacturing into higher-value-added goods. The potential exists for even greater cross-fertilization in the manufacturing industries, as the demand for technology exports expands. Forty per cent of Japan's total technology exports go to East Asian nations, which in some cases rely on Japan for more than half of their imported technology.[12]

The Pacific Basin has an extraordinary wealth of natural resources. The combined tin reserves of Indonesia, Malaysia and Thailand amount to 43 per cent of the world total. The Philippines has 13 per cent of the world's cobalt.[13] Coal reserves, including those of Australia and China, are 34 per cent of the total world reserves.

Even in petroleum, most countries are trying to decrease dependence on Middle Eastern oil and increase imports of oil, natural gas and coal from the Pacific. Malaysia and Indonesia together supply close on 2 million barrels of crude oil per day, and Indonesia has an estimated reserve of 11 billion barrels of untapped domestic resources, with a further reserve of some 29 billion elsewhere in the region. Japan now gets 19 per cent of its total oil imports from East Asia, compared with 7 per cent in 1972. The region is by no means self-sufficient or ever likely to be, but it does have ample supplies to provide a significant proportion of individual country needs.

With Japan emerging as one of the major exporters of long-term capital and direct investment, the Pacific Basin is gradually developing more finance capital independence. Although, in general, the commercial banks are cautious about making balance-of-payments loans to developing countries, in part because of the growing debt crisis in Eastern Europe and Latin America, they continue to regard the Pacific countries, especially those with little external debt, with favour. Also, the latest trend of the banks to increase trade financing to replace other sources of income is particularly well adapted to the current needs of the Pacific countries.[14]

The healthy condition of the Pacific Basin economy has given further stimulus to the proposal for some form of wider economic cooperation which, by reinforcing the area's own trend of growth, would shield it from the ills in other regions of the world. One of the earliest manifestations of the idea was a scheme for a Pacific Free Trade Area (PAFTA), outlined by the Japanese economist Kojima Kiyoshi in 1967, which was intended partly as a counterweight to the growth of the EEC. Although the scheme itself never got under way, it did lead to the establishment of a forum known as the Pacific Trade and Development Conference (PAFTAD), whose meetings were attended by economists from a number of Pacific nations, including the United States, Japan, Australia, South Korea and the ASEAN countries. In 1979, the US Senate Foreign Relations Committee published a report recommending that an Organization of Pacific Trade and Development (OPTAD), patterned after the OECD, be set up; and in the same year the Pacific Basin Cooperation Study Group, organized by Prime Minister Ohira Masayoshi of Japan as part of his search for a conceptual framework for Japan's policies for the 1980s and beyond, submitted a report to the government recommending a further range of cooperative schemes.[15]

The most important development to date, however, has been the establishment of the Pacific Economic Cooperation Conference (PECC) — also called the Pacific Community Seminar. Its first meeting was held in Canberra in September 1980, with a delegation from the five developed countries, South Korea and the five ASEAN countries. Each delegation had three representatives, one from the private sector, one from government and one from the academic world — a 'tripartite' form of representation which was to become standard practice in subsequent meetings. Papua New Guinea, Fiji and Tonga were represented by a joint delegation, and various international organizations sent representatives. The conference recommended that a standing committee be established to coordinate information exchange as well as to set up task forces to study specific issues relating to the problems of economic cooperation.

These recommendations were not immediately implemented, partly because of the need to obtain official governmental endorsement, but, more importantly, because of reservations expressed by some of the ASEAN delegates, who feared that the formation of a larger group of this kind would dilute ASEAN cohesion, and that they would be unduly dominated by the more powerful members, such as the United States and Japan. Also, there was a fear that, by joining a group led by the United States, ASEAN would compromise its traditional stance of neutrality, as well as its position in North-South relations.

However, by the time the second PECC was held in Bangkok in June 1982, ASEAN had managed to agree on a common stand, and was able to state that 'the economies of ASEAN are closely linked to the economies of other Pacific Basin countries, and that economic cooperation with these countries would enhance the economic progress of ASEAN.'[16]

The Standing Committee that was finally established by the Bangkok conference organized four task forces, which were commissioned to study, respectively, trade in manufactures, trade in agricultural products, trade in minerals, and direct investment and technology transfer. The conference decided that the first step in Pacific Economic Cooperation was to hold a series of consultative meetings of the tripartite delegations, whose task it would be to review matters of common concern, and to pass on recommendations to the governments and to the relevant international organizations.

Clearly, the great cultural and political diversity of the Pacific Basin countries will prevent them from forming a tight-knit economic group such as the EEC. The problem of membership alone, which involves such intractable issues as China-Taiwan representation, will defy easy solution. In this regard, the PECC seems to be emulating the wisdom and experience of Southeast Asian regionalism, which skirts the difficult issues and proceeds with what — in any given circumstances — it *can* deliver. The third PECC was held in Bali in November 1983, and the fourth is scheduled to take place in South Korea a year or so later.

Japan's Involvement in the Regional Economy

Japan's GNP in 1982 was equivalent to well over three times the combined GNPs of the five ASEAN countries, Hong Kong, South Korea and Taiwan (see Table 7.1). Because of the size of the economy and its seemingly voracious consumption of raw materials, light industrial goods and, recently, increasingly sophisticated manufactured goods, the heavings of the Japanese economy are felt throughout the region. The fluctuations of Japanese business cycles, movements in the exchange rate, particularly the yen/dollar rate, changes in government policy, especially decisions affecting interest rates, tariff levels and import quotas, have an impact on the economic health of every country. The flow of Japanese capital through both direct and indirect investment mobilizes large amounts of local capital and resources.

Japan's centrifugal impact on the region is inevitable in view of the size of its economy, but this is not always beneficial to the less-developed nations in the area. It is important, therefore, that Japan should realize that dependence on it is one of the principal features of the region's economy, and should try to steer its course in such a way that its activities contribute to greater regional growth and stability.

* * *

Japan is the single largest trading partner for Indonesia, Malaysia, Singapore and Thailand, and the second largest for Hong Kong, South Korea, the Philippines and Taiwan. In spite of successful efforts in recent years to diversify their exports from primary commodities to manufactured goods, the four developing nations of ASEAN, namely Indonesia, Malaysia, the Philippines and

Thailand, are still dependent on one or two commodity exports for the bulk of their foreign exchange earnings: oil in the cases of Indonesia (59 per cent of total exports) and Malaysia (26 per cent of total exports); tin and rice in the case of Thailand (24 per cent of total exports); and copper in the case of the Philippines (8 per cent of total exports). For each, Japan is the largest or second largest purchaser of the commodity, and therefore the continuity of demand and the volume of Japanese purchases is a key factor in determining the balance of payments and foreign exchange earnings, and an important factor in determining levels of industrial activity as well as employment in the relevant sectors.

ASEAN, therefore, has long been eager for Japan to implement some form of price stabilization scheme for exports — a Japan-ASEAN 'Stabex' scheme — providing compensation for losses in export income, and this was officially raised at the Kuala Lumpur summit in 1977. However, Prime Minister Fukuda was against the idea, on the grounds that it would lead to the formation of economic blocs in the world. By way of explanation, he related in some detail his personal recollections of the world financial and economic crises of 1929–31.[17] The idea lay dormant for a while, but was raised again during Prime Minister Nakasone's visit to ASEAN countries in May 1983. The Japanese government is still reluctant to enter into any kind of Stabex arrangement with ASEAN alone, since it hopes to participate in a form of global Stabex with the United States and the European countries.[18] However, in view of the Western countries' lack of enthusiasm for such a scheme (although the EEC countries have their own Stabex with certain developing countries under the terms of the Lomé Convention), Japan will probably have to reconsider some sort of Asian Pacific Stabex.

As Table 7.4 shows, many countries in the region, particularly the East Asian NICs, have substantial trade deficits with Japan. In spite of the intra-industry division of labour which is gradually developing with many of these countries, as their industrialization proceeds, their imports from Japan seem only to increase. The accumulated trade deficit of over $200 billion which South Korea recorded in the 1965–80 period in fact became a political problem and has been one of the main sore points in recent Japanese-South Korean relations. Taiwan felt so aggrieved by its unfavourable trade balance with Japan that in February 1982 it instituted an import embargo on about 1,500 Japanese products; the Japanese had to agree to increase their imports of Taiwanese goods before the ban was revoked. Hong Kong has also complained about its unsuccessful attempts to boost exports to Japan and so correct its huge trade deficit. As regards the developing members of ASEAN, although the high level of Japanese imports of oil and natural gas makes the overall ASEAN trade balance with Japan favourable, individual countries, notably Thailand, have continuously experienced large deficits, a source of serious grievance in an otherwise comfortable relationship.

Table. 7.4: Japan's Trade with the Asian Pacific Region (US $m)

	Japan's exports			Japan's imports		
	1980	1982	%	1980	1982	%
S. Korea	5,368	4,881	3.5	2,996	3,254	2.5
Taiwan	5,146	4,255	3.1	2,293	2,443	1.9
Hong Kong	4,761	4,718	3.4	569	622	0.5
Indonesia	3,458	4,261	3.1	13,167	12,005	9.1
Malaysia	2,061	2,502	1.8	3,471	3,010	2.3
Philippines	1,683	1,803	1.3	1,951	1,576	1.4
Singapore	3,911	4,373	3.1	1,507	1,826	1.2
Thailand	1,917	1,907	1.4	1,119	1,041	0.8
China	5,078	3,511	2.5	4,323	5,352	4.1

Note: Percentage column represents share of Japan's total exports and imports.
Source: Keizai Koho Center, *Japan 1982* and *Japan 1983* (Tokyo, 1982 and 1983).

Obviously, one of the causes of the trade imbalance is the low level of Japan's imports of manufactured goods (about 21 per cent of total imports). This stems partly from the country's basic economic structure, which makes it necessary for Japan to maintain its high level of exports of manufactured goods in order to pay for its ever-increasing import of energy and industrial raw materials, and partly from its geographical and historical circumstances, which predispose it to a go-it-alone style of economy. Until very recently, Japan's consumers were made to understand that it was unpatriotic to covet foreign goods, and, similarly, individual industries generally turned first to domestic suppliers. As far as imports from the East and Southeast Asian countries are concerned, although the percentage of manufactured products has risen from 15 per cent in 1970 to 26 per cent in 1981, import levels for manufactured and semi-manufactured goods need to be much higher if Japan is to help promote the industrialization of the region.

In April 1983, the Nakasone administration announced a fairly comprehensive set of market-opening measures which, it was hoped, would go some way towards correcting trade imbalances. The Asian Pacific countries, however, felt that these new measures were designed to satisfy the demands of Europe and the United States, and did little to help the Asian countries. Therefore, on the occasion of his May 1983 ASEAN trip, Nakasone had to promise to widen the GSP (Generalized System of Preferences) arrangements on light industrial products from the region by 50 per cent.

In the context of Japan's overseas trade, mention should be made of that unique institution, the *sogo shosha* (general trading companies), which are active in the region both in trading operations and in direct investment. The nine ubiquitous major *sogo shosha*, known for their ability to deal with 'everything from noodles to missiles', handle more than half of Japan's total foreign trade (in 1982, 52 per cent of exports and 63 per cent of imports). As far as trade with the Asian Pacific region is concerned, this share is likely to be even greater, especially for imports, since the companies are particularly involved in securing resources of the kind in which the region abounds. They are also widely involved in third-country trade; for example, one *sogo shosha* representative admitted that bilateral Japan-Hong Kong trade accounted for only 25 per cent of his company's Hong Kong turnover.[19] UNCTAD figures suggest that in 1979 the combined textile sales of the nine *sogo shosha* were three times greater than Japan's total imports and exports of textiles, implying that these companies were involved in trading textiles from other countries, probably primarily South Korea and Taiwan, to the Western markets.

* * *

Japan's direct foreign investment (DFI) in the Asian Pacific region was modest until the mid-1970s. In the second half of the decade, however, with the lifting of restrictions on capital outflows (through successive stages of liberalization of capital markets by the Ministry of Finance), the amount increased steadily, so that Japan has now overtaken the United States as the main source of direct investment for the region. As of March 1983, the cumulative total of Japanese DFI in the eight main non-communist countries (see Table 7.5) was $14 billion, approximately one-quarter of total Japanese DFI (and just marginally lower than Japanese DFI in North America).

Japan's direct investment was motivated by three considerations: access to raw materials, relocation of production to take advantage of lower local costs, and the establishment of bases to support trade through banking, insurance, etc. Industrial policy encouraged, in particular, the transfer of industries which were losing comparative advantage in world markets. Rather under half of such industries were in the manufacturing sector, with manufacturing in iron and non-ferrous metals predominating. Mining accounted for nearly a third of the total figure. As Table 7.5 shows, by far the largest East Asian recipient of Japanese DFI has been Indonesia, mainly in the area of energy, oil and natural gas development; indeed it ranks second only to the United States in attracting Japanese investment.

Table 7.5: Japanese DFI in the Asian Pacific Region (Cumulative total, 1951 to 31 March 1983)

	No. of cases	Amount (US $m)
Indonesia	1,148	7,268
Hong Kong	2,002	1,825
Singapore	1,373	1,383
South Korea	1,105	1,312
Malaysia	720	764
Philippines	583	721
Thailand	853	521
Taiwan	1,225	479
Others	335	280
Asia total	9,344	14,553

Source: Japan, Ministry of International Trade and Industry, *News from MITI*, 18 July 1983.

In contrast with the typical US investor, the majority of Japanese investors are middle- or small-scale companies, and most of them tend to invest in labour-intensive industries in the region. There is often participation by *sogo shosha* in the ventures, as one of the partners, since they can make a valuable contribution in terms of marketing know-how, as well as by virtue of their massive financial capability. Whereas American investors prefer 100 per cent ownership, the Japanese have generally made joint-venture arrangements, sometimes even with minority Japanese ownership. This may reflect a lack of confidence on the part of Japanese investors to operate in foreign countries with alien cultures and traditions, but it could also be in reaction to the conditions governing joint ventures with foreign companies in Japan, for which until recently the government has insisted on majority Japanese ownership.[20]

There is little doubt that Japan's DFI is playing a catalytic role in the development of many industries in the region. It accelerated the process of the regional division of labour in many sectors, and provided capital support for the development of the necessary infrastructure in resource-related projects. However, because DFI involves human contacts, it tends to cause socio-cultural tensions and to breed emotional grievances. As a result, one often encounters contradictory criticisms: on the one hand Japanese capital is accused of dominating the local economy, while on the other it is criticized for being too small in amount to play a substantial role in the capitalization, and hence industrialization, of the host country. Although there is little reliable data on this particular point, it seems likely that, in many developing countries, the effect of Japanese DFI in triggering domestic investment through joint ventures is more important than the statistical data suggest.

In the same vein, Japan's DFI is often criticized for not creating enough employment, with very low figures being cited for the proportion of locally recruited employees in Japanese-financed firms as against a country's total workforce. Examples of such percentage shares are 0.1 per cent for Indonesia, 0.3 per cent for Thailand and 0.7 per cent for South Korea. Only in Singapore and Hong Kong are the figures — at 3.8 per cent and 1.9 per cent, respectively — somewhat more respectable.[21] However, given the fact that these figures include the agricultural sector, which accounts for a huge proportion of the total workforce in large developing countries, the net effect of Japanese DFI on employment in other sectors may well be underestimated.

There are also complaints about insufficient technology transfer on the part of Japanese firms — a rather serious charge, since the transfer and dissemination of technology and managerial expertise to the host country are perhaps the most coveted aspects of DFI. Here again, however, the data are ambiguous and contradictory. Many Japanese firms claim that they provide both on-the-job training and courses, lasting several months, at headquarters and factories in Japan. They say, moreover, that it is they who have cause for complaint, since many employees thus trained leave the company upon completion of the training period. These problems, stemming from socio-cultural differences, are precisely those that call most urgently for comprehensive study — perhaps at government level. It is good to know that the PECC has commissioned one of its task forces to study investment and technology transfer in the Pacific Basin countries, because in this particular area it is the empirical information which is most lacking.[22]

In the coming years, as regional interdependence grows, the Asian NICs are likely to emerge as the new investors. Singapore was already the second largest foreign investor in Malaysia as of July 1980, while Malaysia has begun investing in Thailand and other countries. There is every reason to believe that South Korea, Taiwan and Hong Kong will follow suit and will emerge as the new actors in intra-regional DFI, thus diversifying investment sources in the region. Conversely, as Japan's economic growth slows down, its DFI, particularly in resource-related projects, may not continue to increase at its previous pace. Current negotiations on LNG development in the region are a case in point.

After making substantial commitments to the development of natural gas facilities in Brunei, Indonesia and Malaysia throughout the late 1970s, Japan finds itself faced with the likelihood of a surplus in contracted purchases of LNG for the remainder of the decade. Contracted LNG imports from Brunei and Indonesia had already reached more than 14 million tonnes per year by 1981 (a figure that is likely to rise to 30 million tonnes by 1986, the Malaysian project having come on stream in 1983). Thailand, which is seeking Japanese development assistance for its new offshore gasfields with a view to quartering its purchases by 1988, is running up against reluctance on the part of Japan to overextend its purchase agreements or to build up surplus capacity in the region; on his visit in May 1983, Prime Minister Nakasone said that Japan would be unlikely to be able to import Thai LNG before the end of the decade.

* * *

In 1977, under Prime Minister Fukuda, Japan promised to double its Official Development Assistance (ODA) and to raise the level of ODA as a percentage of GNP from 0.21 per cent to 0.32 per cent by 1980 — a target that it accomplished even earlier than promised. This was partly in response to criticism of its relatively poor performance as an aid donor, on the grounds (1) that it failed to meet the ODA target of 0.7 per cent of GNP, (2) that the grant element of aid averaged 74 per cent as against the 80 per cent norm set by the OECD's Development Assistance Committee (DAC), and (3) that about two-thirds of its aid was tied to purchases from Japan.[23]

But Japan's efforts to improve its standing as an aid donor were not merely in response to criticism. There is overall agreement in Japan on the need for an expanded commitment to regional growth and stability, and it is widely accepted that Japan's own economic security lies in Pacific prosperity. Japan's uniquely non-confrontational policy towards the Third World, which is due, at least in part, to its heavy reliance on resource supplies, has been shaped with such goals in mind. Pressure, both from fellow DAC countries and from the Group of 77, are further reasons for Japan to fulfil its obligations in contributing to Third World needs.

Within the government, the Foreign Ministry is naturally the most committed to foreign aid objectives, but the Ministry of Finance, too, seems to agree with this general policy line, as was shown by the fact that the ODA element of the budget was allowed to rise 8.9 per cent in the otherwise austere budget of 1983 (defence being the only other element to obtain an increase over the previous year). However, given the overall slowing down of Japan's economy (the Economic Planning Agency forecasts a growth rate of no more than 4 per cent or so for the remainder of the decade), the government will find it difficult to increase its contribution to the region, either in aid or in government finance. In fact, the total flow of economic cooperation funds, including official loans to international organizations, recorded some decline in 1979 and 1980.

The tax system, which was designed at the time of high growth and high inflation, is finding it hard to generate adequate revenues in the low-growth/low-inflation conditions of the 1980s — a state of affairs which is causing regular budgetary deficiencies. The rapidly aging population (since 1950, Japan has had the world's most rapid rise in longevity) is bound to put more strain on scarce government resources, besides its impact on the labour market. In spite of promises from both the Suzuki and the Nakasone administrations to restructure government finances, the practice since the mid-1970s of paying for deficits in the budget by issuing bonds has left the government with yawning debts that neither prime minister has been able to sort out — nor is any immediate successor likely to be able to do so. The total outstanding debt now amounts to about one-third of GNP, and the annual allocation of funds to service the debt has now reached one-quarter of the total budget.

The implications of all this for the regional economy are clear. One likely casualty is Japan's target of doubling its ODA again by 1986. In order to meet this target, the ODA element in the General Account would have to grow by nearly 14 per cent in 1984 and 1985. Since the Ministry of Finance anticipates an annual rise in overall expenditure of only 6.8 per cent in these years, it is extremely unlikely that aid expenditure will increase at double the rate of other budget items.

There is no doubt, however, in view of the size and scope of its activities, that Japan will remain a determining factor in the performance of the regional economy during the 1980s and beyond. The reliance on Japanese capital for massive development, especially resource-processing and large-scale industrial projects, is bound to grow, if only because of the widespread 'aid fatigue' of other developed countries. The fact that the Asian Development Bank recently tried to raise the proportion of Japanese funding is a clear indication that Japan will be asked to carry increasing financial responsibility for the region.

Chapter Eight

JAPAN'S PLACE IN THE WORLD

There is little doubt about the importance of the role which Japan will play in the Asian Pacific region in the 1980s and beyond. As a dominant trade partner for almost all the countries in the region, as well as a major source of aid, finance and technology, its presence is already one of the vital determinants of the region's future. The scope and depth of its current relations with the region are a far cry from what they were a decade ago. Since Prime Minister Fukuda's August 1977 visit, ASEAN countries and Japan have come to understand a lot more about each other through steadily increasing channels of communication, and have learnt to manage their relations intelligently and effectively. Successive Japanese prime ministers have made it a rule to visit ASEAN countries as soon as possible after taking office, and Japanese foreign ministers are regular participants at ASEAN's expanded ministerial conferences.

The November 1983 visit to Japan by Party Secretary Hu Yaobang was the latest in a series of visits by Chinese leaders — Deng Xiaoping in 1978, Hua Guofeng in 1980, Zhao Jiyang in 1982 — marking an increasingly close relationship between the two countries. In talks with the Japanese government, Hu pledged that China would not back down from its open economy orientation; stated that China had such complete trust in Japan's political intentions that it would not object to an increase, even if substantial, in Japan's military capabilities; and requested Japan's continuing cooperation in the modernization of the Chinese economy, which was expected to show a growth rate of 9 per cent in 1983.[1] If this rate of growth continues, and if the country's leadership remains so astute, China might indeed achieve its ambitious goal of a fourfold expansion of the economy before the end of the century. In view of China's size and its very special place in the international system, not only would this have a tremendous impact upon the region's economy, but it would have far-reaching implications for the world as a whole. Needless to say, Japan is a decisive factor in this eventuality.

Japan, however, despite its undisputed economic stature, still has difficulty in establishing its identity and coordinating its policies under a plausible set of goals. Many are mystified by its tendency to avoid political involvement either in the region or in the world. It appears to lack a clear sense of purpose, and as a result it is widely criticized for its indecisiveness in foreign policy and its seemingly beggar-my-neighbour economic behaviour. As for defence, it arouses ambivalent feelings in its partners: on the one hand, its Asian neighbours are wary of its re-emergence as a military power; on the other, the United States is critical of the low level of its defence expenditure, and hence of its perceived 'free riding'. This last chapter approaches the question of Japan's place in the world by looking, first, at its unique sense of isolation and dependence; second, at the way it defines its security problems; and, third, at its regional relations, and hopes in this way to provide some basis on which to speculate about the future direction which the country is likely to take.

Isolation

As was noted in Chapter 1, there has been little similarity or coordination between Japan's process of development and that of the region. In fact, after the war, Japan was literally banished from continental Asia, and had to survive virtually on its own resources, with little prospect of being able to take part in any regional system that might develop. With wartime memories still fresh, few countries in either Northeast or Southeast Asia wanted its involvement in the region in any form or context. Whatever interaction was beginning to take place between Japan and the region before the war, notably with Taiwan, Korea and Manchuria, had been cut off completely, and it seemed unlikely that any kind of division of labour would develop in the near future.

As a result, Japan had to survive as an isolated entity in the post-war world, without any kind of international or regional support system. Even after the signing of San Francisco Peace Treaty, it was able to claim little status internationally, except that of *de facto* protégé of the United States, which obviously needed Japan as its strategic base in East Asia. Few other countries had any reason to support Japan's cause, whether in its relations with Europe, the treaty with the Soviet Union or its complex interaction with China.

For a long time, the favour and support of the United States were the only bargaining counters that Japan could use in its dealings with the world.

This was very different from the case of West Germany, which was included in the emerging West European system right from its birth as a nation. At the end of the war, Germany was divided into four zones and occupied by four different countries, with the understanding — reached at the Potsdam Conference in July 1945 — that the four zones would eventually be treated as one economic unit. The Soviet Union, however, soon made it clear that it intended to keep its zone within the communist sphere of influence, and the Allied countries responded by merging their zones and founding the Federal Republic of Germany — West Germany. The Russian response was to proclaim the Democratic Republic of Germany — East Germany. In other words, West Germany was conceived and founded as a part of the emerging system of Western Europe, in the context of the post-war confrontation between communism and the free world.[2]

As was shown by the results of October 1949 elections for the Bundestag, the West German people supported the proposal. They gave their agreement to Chancellor Adenauer's contention that the best means of ensuring their survival as a nation and regaining their status in the community of nations would be to allow the new republic to be fully integrated into the emerging system of Western Europe and to participate in its construction.[3]

The three zones that went to make up the new republic all lay in the western part of the country, centring on the Rhineland industrial region. Unlike pre-war Germany, therefore, which extended its influence east and west, the Federal Republic was firmly anchored in the West. Historically and culturally, its links were with the western half of Europe, and this, too, helped the process of political integration. Economically, as well, West Germany was only half an entity in that it had lost the important agricultural areas east of the Elbe on which pre-war Germany had relied. Furthermore, the entire system of distribution, communication and transportation services had been cut in half. In other words, the West German economy could not have functioned without integration.[4]

Thus, West Germany soon found itself becoming an important component in an extensive division of labour that was developing throughout Western Europe.[5] Moreover, thanks to clever economic management and the healthy pattern of growth that resulted, it soon came to be recognized not only as a valuable contributor to the West European economy, but also as a ready source of aid and assistance to its neighbours.

In contrast, the stigma of isolation remained with Japan, being reflected both in patterns of development inherited from the past and in its current economic behaviour. In order to reconstruct the economy, which had been devastated by the long war of attrition, and to provide the population with the basic minimum of food and work, priority was given to reviving whatever portions of the pre-war economic system still functioned. In the face of acute food shortages, even starvation, there was no alternative but to give generous protection and subsidy to the existing agricultural section, regardless of cost-efficiency considerations. Even in other sectors, it was not until after the 1960s that the concept of international integration began to appear, even in its most rudimentary form; until then the prevailing pattern was to manufacture everything domestically, and to export as much as possible in order to pay for the carefully selected and severely restricted imports of vital natural resources. These are precisely the causes of many of the difficulties which Japan was to face later, such as the small percentage of manufactured goods in its total imports — a constant source of friction between Japan and both developed and developing countries — and its reluctance to increase agricultural imports, which is a cause of one of the main strains in current US–Japanese relations.

It seems likely that the realization of Japan's continuing isolation, and the difficulties it entailed, was one of the factors which encouraged Prime Minister Ikeda to adopt an apolitical, economistic stance in the early 1960s. Certainly there seemed to be no way for Japan to involve itself in the politics of the region, which were becoming increasingly confrontational, and it was obviously difficult to stay without some kind of a purpose and still hope to strengthen its position in the world. Ikeda's contention that Japan should establish its international status through economic development touched a chord with the people, who, after a decade of sterile ideological polarization in the country, sensed the depth of Japan's international isolation, and were troubled by it. Ikeda's call rallied them to the immediate and practical goal of development, and plunged the nation into the high-growth decade of the 1960s.

By the 1970s, however, the Japanese had discovered, to their dismay, that economic power alone does not necessarily lead to international status and respectability. The series of shocks which they experienced during the 1970s drove home to them that economic power without adequate political support can make a nation more, rather than less, vulnerable. With a mixture of awe and envy, they watched West Germany, which seemed to be able to navigate the difficult seas of international politics with all the composure of a mature nation, launching such important new policies as Ostpolitik, or dealing brilliantly with such events as the Arab hijacking of the Lufthansa aircraft at Mogadishu airport. (This incident stood in sharp contrast with Japan's total capitulation to its own Red Army hijackers at Dacca airport in Bangladesh at about the same time.) What they did not realize was that there was a price tag attached to acquiring an international position, and that the price paid by West Germany had been no less than the loss of the old entity that was Germany.

Pre-war Germany used to wield a decisive influence on the affairs of Europe. Not unlike Japan in East Asia, it too was prone to be the source of regional instability, as shown in the three major wars that it fought in a period of less than a century. The division of Germany, therefore, must have been a welcome development for the rest of Europe. Although it had been brought about by the cold war, at least Germany's neighbours could obtain some measure of security from the knowledge that a divided Germany was, by definition, a less aggressive Germany. They allowed West Germany to share in their collective position and status, without any great fear of its threatening their peace again.

Similarly, at the end of the war, something had to be done to neutralize Japan's destabilizing potential, either by dividing it or by isolating it from the rest of the region. In the event, on account of the Americans' determination not to let the Russians come into East Asia, Japan was placed, undivided, under the single authority of the United States. As a result, it retained its pre-war entity intact, although it was thoroughly disarmed, with all its colonial possessions taken away. Even then, many nations in the region were uneasy about this arrangement, as shown in some of their reactions to the San Francisco Peace Treaty. Although they finally accepted the treaty, under strong American pressure, they did so only on the understanding that the United States would keep a tight rein over Japan's activities for the foreseeable future. Perhaps it was this background that produced the peculiar tentativeness of Japan's post-war stance and, subsequently, its excessively dependent relationship with the United States.

Dependence

In March 1982, Wakasugi Kazuo, the director of MITI's Trade Policy Bureau, told the press that 'if the United States and Europe do not trade with Japan, there would be no benefit for Japan to remain a member of the free world. If that happens, we would probably join the communist bloc.'[6] Predictably, this statement caused some stir around the world, although Wakasugi hastened to add that he was speaking only hypothetically. Another suggestion, made by an Indian scholar, was that Japan should join the non-aligned movement. He pointed out that, on account of its 'peace constitution', as well as its relatively clean record as regards, for example, Islam and black Africa, it was eminently qualified for membership. He contended that the movement would thus be greatly strengthened, while Japan would find the goal it had been looking for — to work for the peace and freedom of all mankind.[7]

Clearly, neither of these are realistic options. Japan has to be part of the free world by virtue of the size and structure of its economy. Moreover, it operates its economy, the only basis of whatever influence it has in the world, under conditions of great vulnerability, and it depends entirely on the United States for political support and a military defence guarantee. Even the slightest sign of a rupture in the relationship with the United States would plummet the value of the yen and undermine confidence in Japan's ultimate viability; and this, in turn, would cause a disastrous setback to the prospects of the region as a whole.

Also, by now, the relationship with the United States has grown too big for Japan to be able to tamper with it for the sake of some temporary advantage. In 1982, the United States took $36.3 billion worth of Japanese goods, or 26.2 per cent of Japan's total annual exports, and Japan was the biggest single importer of American agricultural products, with the total imports from the United States amounting to $24.2 billion, or 18.3 per cent of Japan's imports. Together, the United States and Japan account for roughly one half of the free world's GNP, and are vital to the functioning of the entire world economy. Although Japan may not always see eye to eye with the United States on its cold-war policy, on almost every other issue the two countries are basically in agreement. Added to this, with a substantial and constantly growing exchange of people and ideas, the two countries have forged what US Ambassador Mike

Mansfield has called 'the most important single bilateral relationship, bar none'.[8]

Nevertheless, there is a peculiar love-hate syndrome in US-Japanese relations, which comes to the surface from time to time and puts considerable strain on the policy-makers of both countries. Its typical pattern is that, when Japan is small and weak, the United States is kind and generous, willing to extend every kind of assistance, and when Japan becomes big and strong, the United States gets annoyed and resorts to Japan-bashing.

That is what happened in the early part of this century when Japan's victory over Tsarist Russia led the United States to make the thwarting of Japan the centre-piece of its Asian policy. Admittedly, Japan deserved the 'bashing' for its blatantly militaristic and colonialist behaviour, but American criticism went beyond justified indignation and took on a self-righteous and evangelistic character — an attitude that was clearly counter-productive. Likewise, after World War II, when Japan was weak, the United States went out of its way to help it. More than that, it became a combination of guardian-angel and reformer, taking complete responsibility for every aspect of its rehabilitation and development. It was an extraordinary relationship, but one that was well attuned to the psychological needs of both countries at that particular time. Japan was defeated, humiliated, insecure; the United States was triumphant, confident, generous, and ready to lead and protect its weaker ally.[9]

The honeymoon of dependence lasted long after the Occupation was over. Japan's rapid economic growth, coupled with the two countries' increasing divergence of view on the cold war in Asia, brought into play the historical pattern of relations. The momentous shift of US Asian policy at the end of the 1960s accelerated the process. Frustrated by growing trade deficits with Japan, annoyed by the unexpected inroads of Japanese exports into politically sensitive markets and resentful of Japan's perceived 'free riding', the United States once again adopted the practice of Japan-bashing, this time euphemistically calling it pressure tactics, in order to try to correct the glaring imbalance of the relationship.

In the contrast with pre-war days, Japan in the 1970s was no longer in a position to oppose the United States, let alone to go to war with it. In part because of its economic success, its dependence had increased enormously. It had no alternative but to try to accommodate US demands, although this was not easy, partly because of the peculiarity of Japan's post-war economic structure, and partly because American demands did not always appear to be economically sound, and seemed at times to be tantamount to punishing Japan for its hard work and technological excellence.

To return to the analogy with West Germany: although the relationship with the United States was also of primary importance, West Germany was able to introduce some balance into it by means of the dialogue with other members of the Western alliance. The opportunity to discuss the import of some proposition or demand made by the United States with countries of similar size and environment, and to evaluate collectively the implication of a prospective response, must be of great value to Germany in the management of its most important relationship.

* * *

In this connection, mention should be made of the concept of trilateralism — the doctrine advanced by the Trilateral Commission, an organization that was launched in 1973. The concept was welcomed by Japan as a possible means of strengthening its position, particularly vis-à-vis the United States. The idea grew out of the perceived need for the three 'industrialized democracies', namely the United States, Western Europe and Japan, to join together in dealing with problems that affected all their regions, such as energy security, North-South questions and East-West trade. Also, it was seen as a way of including Japan in the discussions of the advanced industrialized countries — the need for which was clear in view of its importance both as a trade partner and as a procurer of the world's natural resources.

For Japan, a multilateral dialogue was certainly a chance to help balance its relationship with the United States. Also, from the point of view of trade relations, it would certainly make sense, since Japan was beginning to feel more and more like a lonely outpost of the industrial world in Asia.[10] Finally, to be invited by the Western powers as an equal partner was gratifying in itself.

However, trilateralism could not provide the answer to all the difficulties regarding its position and status in the world. For instance, in view of its growing involvement with the countries in the Asian Pacific region, it would be awkward if Japan were to give the impression that it was choosing the West as its principal partner, or to identify itself strongly with the wealthy developed North. With that in mind, when invited to attend the seven-power economic summit at Rambouillet in December 1975, the first such meeting to be held, Prime Minister Miki took care to send a special envoy to the ASEAN countries to explain Japan's attendance at the meeting. He did this partly out of embarrassment at being the only Asian at such a high-level Western gathering, and partly because he did not want to jeopardize Japan's chance to play a role in facilitating relations between Asia and the West — an aspiration which the Japanese have long cherished, although little of substance has yet emerged.

Also, in its sweeping generalization, the concept of trilateralism had a tendency to overlook differences in structure as well as in role of each of the three regions. There is little doubt that, in terms of US defence strategy, Western Europe and Japan were the areas which must be defended at all costs, and that Western Europe and Japan have formulated their policies on a US defence guarantee. However, the threat which the Soviet Union is imposing on Western Europe is somewhat different in character from that which Japan perceives as a threat to itself and to the Asian Pacific region. The next section, which examines Japan's perception of its security needs, may help to clarify some of these points.

Security

The fact that Japan is a group of small islands, separated from the Asian continent by a strip of water that is 200 km wide at the nearest point, seems to have two contradictory implications for the nation's security. On the one hand, it emphasizes its extreme vulnerability. A huge population on a group of small islands, with a dense concentration of industry and communication systems in the narrow stretch of land that lies between Tokyo and Osaka makes Japan virtually impossible to defend. The Japanese in fact agree with the view, allegedly held by the Russians, that 'one bomb is enough to annihilate Japan'. This situation, coupled with memories of World War II and the holocaust at Hiroshima and Nagasaki, has effectively destroyed whatever faith the Japanese had in the ability of military means, conventional or nuclear, to protect their national interests. For ordinary Japanese, war is a non-option. The 'peace constitution', with its celebrated Article 9 renouncing war, gives this attitude international legitimacy.

On the other hand, the Japanese are also aware that the narrow strip of water which separates their islands from the continent can work as an effective deterrent against any attempt at invasion. After all, Japan was never seriously attacked throughout its long history, except in 1274 and 1281 when the Mongolians sent their ships to the island of Kyushu. On both occasions, the Mongolian fleet was hit by a typhoon (*kamikaze*) and sank in the Sea of Japan. Even the United States chose to bomb Hiroshima and Nagasaki rather than risk landing operations, which it reckoned to be too costly. In other words, although an enemy might find it easy to destroy Japan from the air, there would be considerable hazards in trying to land ground forces. Only a substantial force could negotiate such an operation, and this would give the United States time to come to Japan's aid. Therefore, unlike Western Europe, which must resort to theatre nuclear forces in order to make up for the deficiency in its conventional armament, Japan can still conceive a credible non-nuclear option for deterrence.[11]

As a superpower in a deadly tug of war with another superpower, the United States regards the Soviet Union as the common enemy of the entire free world. It believes that other nations should think so too, and should therefore help cover the cost of the US arms build-up. In US terminology, this is 'burden-sharing', and it forms the basis of American criticism of Japan's so-called 'free ride'.[12] For Japan, however, the Soviet Union is not the main threat. It is a dangerous adversary, but one with considerable weaknesses. Nor does Japan share the American view that the Soviet Union is irrevocably committed to aggression in East and Southeast Asia.

What is as threatening to Japan as invasion by the Soviet Union is a disruption of energy and food supply. Having to sustain 118 million people on four small islands, an area not quite the size of the state of California, 85 per cent of which consists of forbiddingly mountainous terrain, Japan is deeply obsessed by the problem of energy and food security. The United States would have little difficulty in living off its own resources. Western Europe, too, has far more substantial natural resources than Japan, and in any case the West European nations can exert considerable collective influence and bargaining power, and can close ranks and act together against a possible disruption in supply. Thus, the concept of 'comprehensive security' takes on a pressing urgency in the minds of the Japanese.[13]

The Task Force Report on 'Comprehensive National Security', sponsored by Prime Minister Ohira's administration and submitted to it in July 1980 shortly after his untimely death, is an examination both of Japan's position in the world and of its vulnerability as perceived by the Japanese as they entered the 1980s. The whole concept of 'comprehensive security' is based on the no-war premise of Japan, and therefore emphasis is placed on maintaining the peace and stability of the world, preserving the free trade system and securing access to vital natural resources, rather than on the purely military aspect of defence. Admittedly, it makes extensive recommendations on how and why the country should try to improve its defence capability, in accordance with the on-going defence programme approved by the government in 1976, but the emphasis is on the prevention of an enemy landing. Moreover, even the hypothetical case of an invasion, as well as Japan's response to it, is treated not so much as war as crisis management. Thus, in addition to enhanced military preparedness, the recommendation calls for an improvement in the nation's capability to cope with a variety of crises, including earthquakes.[14]

The Ohira administration had commissioned the study on 'comprehensive security' at the end of 1979, when it had to cope with the Soviet invasion of Afghanistan and the subsequent changes in US global policy. It complied with the US initiative on the boycott of the Olympic games in Moscow in the summer of 1980, and took a positive action in extending substantial aid to such strategically sensitive countries as Pakistan and Turkey, the latter reportedly after persuasion from Germany.[15] Although fully aware of Japan's dependence on the United States, Ohira was eager to define and articulate Japan's position and role in the world of the 1980s, and initiated a number of studies, including that on the concept of 'comprehensive security'. However, the US administration, increasingly obsessed with the 'Soviet threat', tended to dismiss the whole concept as a rationalization of Japan's 'free-riding'.

During the first three years of the 1980s, the relationship with the United States showed particular strain, in part because of the different style and perceptions of the leaders then in power. Prime Minister Suzuki Zenko gave priority to harmony within his own party, and was therefore painfully indecisive on many important issues, particularly defence; President Reagan, strong and assertive, was firmly committed to a global crusade against communism in general and the Soviet Union in particular.[16] It led to something of a catastrophe in May 1981 when Suzuki, on his return from Washington, denied the 'alliance' with the United States which he had clearly pledged in a joint communiqué with Reagan just before. It was an extraordinary blunder and led the foreign minister, Ito Masayoshi, to resign.

When Nakasone Yasuhiro became prime minister in December

1982, the tension between the two countries was somewhat relieved. Nakasone made no bones about the importance for Japan of the alliance with the United States and clearly stated that Japan should play a bigger role in the world by boosting its defence capabilities.[17] However, even under Nakasone's unusually assertive leadership, it is questionable whether Japan will move much beyond the conceptual framework laid out by the Task Force Report, since even its very moderate recommendations seem to arouse alarm not only in the opposition parties, but among the majority of the population.

Japan has long adhered to a policy of nuclear restraint as well as restraint on the export of arms. Its nuclear policy is embodied in the so-called Three Non-Nuclear Principles, declared national policy by Prime Minister Sato in 1968, which specify that 'Japan will not produce, possess or let others bring in' nuclear weapons. These principles seem to have reassured the countries in the region of Japan's commitment to peace and presumably have helped to reduce tensions in East Asia. Also, for all its 'economistic' orientation, Japan has consistently refrained from participating in the lucrative arms trade. In 1967, Prime Minister Sato disclosed to the Diet certain criteria for discouraging the export of arms; these were reiterated and somewhat expanded in 1976 by Prime Minister Miki; and, in March 1981, a resolution was passed in both houses of the Diet approving these standards. The Japanese are well aware that by abstaining from the weapons trade, they are losing precious bargaining chips with prospective buyers in the Third World, including those in the Middle East. However, there has been little coordinated pressure from the business community to liberalize the weapons trade, which suggests near-unanimity on this point. Coming from Japan, which rarely projects a political philosophy, the consistency with which these principles have been upheld is significant in that they may very well indicate Japan's choice for many years to come.

* * *

At the time the Nakasone administration took office, the United States seems to have been re-evaluating the importance of Japan in its global policy. Although the tripartite balance of power was a brilliant success in its initial stage, the United States eventually had to face its costly repercussions. Alarmed by the imminent possibility of being encircled and indeed 'contained' by the US-China alliance, the Soviet Union launched an all-out military build-up so that it could fight its war on several fronts. Coinciding with the temporary slack in US defence efforts after Vietnam, this led to a perceived gap in the military balance between the two superpowers, which became quite pronounced by the end of the 1970s. The Soviet Union reportedly holds 120 submarines in the region, including 30 nuclear-powered ones, which can hit targets on the West Coast of the United States with missiles launched from the Sea of Okhotsk. On top of this, it is widely recognized that 70 Backfire bombers are stationed in Siberian air bases, not to count the possible deployment of SS-20s east of the Urals.[18] In view of such a threatening posture on the part of the Soviet Union, including its invasion of Afghanistan, the United States began to make a serious effort to redress the imbalance.

When the direct military competition with the Soviet Union came to the fore, the Kissingerian concept of the tripartite balance of power lost some of its validity, and China's perceived role in strengthening the US position vis-à-vis the Soviet Union was considerably reduced. Unlike the other two major powers, China is an underdeveloped country, and as such its role in the global military balance is necessarily limited. Deng Xiaoping himself told Japanese visitors in 1980 that China's role in a prospective showdown with the Soviet Union would have to be a defensive one. He pointed out that, being a poor country, China had less to lose from a Soviet attack than the developed countries, notably Japan; it could therefore resist invading Soviet forces almost indefinitely, by employing guerrilla tactics. This knowledge would make an effective deterrent to the Soviet grand design in Asia, and that, according to Deng, would be China's contribution in the global anti-Soviet alliance which he was eagerly peddling at that time.[19]

Norman Bailey, director of planning at the National Security Council, in a speech made at the International Monetary Conference in Brussels in May 1983, stated that, in view of the military vacuum in the Pacific basin, where 'a thin array of American forces and a huge but largely impotent Chinese army' are facing the enormous forces of the Soviet Union, it would obviously be advisable to fill the vacuum 'through close cooperation with the Japanese'. Another White House source is reported to have said that Japan is 'a primary partner [of the United States] on the global stage'.[20]

And it is not only from a security point of view that the United States seems to be taking note of Japan's position at the heart of the Asian Pacific region. Secretary of State George Shultz said in a speech in March 1983 that the dynamism of the region which he had witnessed had convinced him that 'the region had an important future, and the United States should be part of it'.[21] The importance of Japan to the United States in this regional context is more than obvious.

Also, with its economy in a little better shape than those of West European nations, Japan can perhaps cooperate with the United States more positively in its efforts, say, to prevent the mounting debt of the Third World countries from getting out of hand, or to maintain the principle as well as the substance of free trade, which is the life-line of the economies of both the United States and Japan. The way in which Prime Minister Nakasone cooperated with President Reagan at the Williamsburg summit in promoting US initiatives, particularly in getting an endorsement for a new GATT round, indicates that a community of interests is beginning to emerge between the two countries.[22]

If a broader convergence of interests were to develop in the future, the one-sidedness which has plagued the past relationship may be replaced by a more symmetrical framework of burden-sharing. Given the crucial importance of the US relationship for Japan, the development of a constructive relationship of this kind would have far-reaching implications for its position in the world. With cooperation with the United States as a frame of reference, Japan would perhaps be able to define a global role commensurate with its ability and its aspirations. Once that had happened, Western Europe, in turn, would be able to develop a clearer perception of Japan's position and its potential. The countries in the Asian Pacific region, too, would see their relations with Japan in clearer perspective. Above all, a redefinition of this kind, by helping Japan to come to terms with its perceived role in the world, would gradually alter the pattern of Japanese thinking and behaviour.

Regional Relations

Unlike the European countries, Japan does not have in close proximity a group of nations of similar size and aspirations with which to relate. Relations with China have been dominated always by an extreme asymmetry in size, population and level of civilization, resulting in a near-constant one-way traffic of cultural input from China to Japan during most of their fifteen centuries of historical contact, except in the first half of the twentieth century, when Japan tried to dominate the continent, invaded China and ravaged the length and breadth of the country.

After World War II and the establishment of the communist government in 1949, China became one of the star players in world politics, although it was still quite underdeveloped socially and economically. From then until the normalization of relations in 1972, the interaction between the two countries went through frequent ups and downs primarily due to the political upheavals in China; Japan, by and large, remained a passive partner, adjusting itself to the vicissitudes of China's political mood. A channel of private trade had been opened up in 1952, partly thanks to the shrewd pragmatism of the leaders of the two countries, and partly because the Chinese hoped to use trade as a lever of influence over Japan. Although affected to some extent, trade relations were not greatly disturbed by China's internal upheavals, and in fact grew steadily, so that by 1969 total bilateral trade had topped $600 million, making Japan China's biggest trade partner after Hong Kong.

In the post-normalization period, Japan continued to maintain a low posture, throughout the post-Mao convulsions and the rise and fall of the Gang of Four, hoping for reason to prevail. When Deng Xiaoping returned to power and tried to steer China towards the less ideological goal of modernization, Japan was eager to help. By this time, because of its outstanding economic success, Japan saw its role as reinforcing the positive elements in Chinese policy, which it believed would stabilize both China and the region. Bilateral trade increased nearly ten times during this period: from $1.1 billion in 1972 to $10.3 billion in 1981.[23] In the summer of 1978, the Treaty of Peace and Friendship was concluded, with Japan agreeing to include in Article 2 the long-disputed anti-hegemony clause, although a disclaimer had to be written into Article 4 that this would not affect either country's relations with any third country. The treaty represented a decisive tilt towards China in Japan's stance vis-à-vis the two superpowers and a break with 'equidistance' diplomacy.

Deng's visit to Japan in October 1978, for the ratification of the treaty, raised a 'China fever' in Japanese business circles, whose leaders foresaw great potential under Deng's pragmatic leadership. In 1981, however, the 'fever' was somewhat dampened by Beijing's so-called 'adjustment' policy, which led to the cancellation or postponement of a substantial number of export contracts. The Japanese took these abrupt changes with good grace, understanding the difficulties which China was facing in its modernization programme. In fact, China has since demonstrated considerable skill in economic management, showing a healthy growth rate in 1982 and 1983, while its foreign exchange reserve exceeded $10 billion at the end of 1983.

China-Japan relations have never been so pervasive as they are now. The interaction between the two countries — in sports, culture, tourism — is massive, cordial and efficient. However, China is a communist country and it is not likely that a political partnership will develop, at least for the present. This is perhaps just as well. History shows that Japan's over-involvement with China and its colonialist ambitions brought the United States into the region and created the precarious Sino-US-Japanese triangle which eventually led the region to war. In the world of the 1980s, this particular triangle is unlikely to recur because of the far more dangerous prospect of superpower confrontation. However, the potential for such competition is inherent in the geopolitics of the region. Therefore, Japan is perhaps well advised to be sensitive in its relations with its two giant partners.

* * *

Korea is a country which, comparable in size with Japan and having many affinities in culture and language, could have been Japan's most natural partner. In fact, Japan owes much of its cultural heritage to the Korean peninsula, and its people — artists, scholars, craftsmen — who emigrated to Japan over the ages have helped to shape Japanese culture. However, in the nineteenth century, the Yi dynasty was painfully slow to realize that the Korean peninsula was being made into a corridor of confrontation between Tsarist Russia and the newly emerging Japan. Also, it underestimated the threatening character of Japan and allowed it gradually and inexorably to infiltrate the peninsula, on the pretext of the need to pre-empt the Russians' southward thrust. Imitating the operational patterns of Western imperialism and applying them in an even more efficient and ruthless manner, Japan turned Korea into its colony in 1910, thus beginning some thirty-six years of oppression and plunder.[24]

The feelings of guilt and remorse for the past on the part of the Japanese, and the memory of humiliation and depredation by the Japanese on the part of the South Koreans, were the basis on which the two countries had to build their post-war relationship. Unfortunately, the political environment of Northeast Asia after the war was totally different for the two countries and resulted in totally different attitudes and perceptions. Thus, whereas for the Koreans, the Korean war was the most painful experience of their entire history, for the Japanese, whose security was firmly guaranteed by the United States, it was but a fire on the other side of the river.[25] This led to glaringly opposite assumptions as the two countries explored their way through the 1950s and 1960s.

Like West Germany, South Korea is a country formed under cold-war conditions, committed since the Korean war to its position as a frontline state in the defence of the free world. Japan, for its part, doubts the validity of ideological confrontations in post-war Asia, and tends to think of the cold war as, at most, a temporary phenomenon. South Koreans cannot understand this attitude and, in view of the fact that Japan's security and economic development has been made possible by the free-world system, regard it as irresponsible and selfish to the point of being immoral.

As a result, Japan has been inclined to take an even-handed approach to North and South Korea, and to irritate the South by its tendency to play down the threat from the North. In addition, a substantial proportion of Korean residents in Japan retain allegiance to North Korea, making Japan an ideological battleground for the two Koreas. The Mun Se-kwang incident in 1974, in which a Korean resident in Japan sneaked into Seoul and, in trying to assassinate President Pak, killed his wife by mistake, is an example.

The relationship has been further complicated by the South Korean internal situation, and successive governments have accused Japan of interfering in the country's internal affairs. The tight security situation in South Korea tends to make the government highly sensitive to internal dissenting movements; this has not only prevented the development of a genuinely pluralistic society, but has led to embarrassing incidents such as the Kim Dae-jung affair. Japan has been continually critical of such repressive governmental methods, and its left wing in particular has made no attempt to disguise its opposition, with little understanding or sympathy for the difficulties that South Korea has faced. This criticism on the part of the Japanese has irritated the Koreans for two main reasons. First, the Koreans themselves were opposing these excesses of the government, in many cases at the risk of long-term imprisonment or even death, and the criticism from outside was embarrassing. Second, they felt — understandably — that, in view of its previous ruthless domination of Korea, with no consideration for human rights whatever, Japan was hardly in a position to be critical.

To make matters even worse, there has been a perennial trade imbalance between the two countries. Since bilateral trade relations resumed in 1965, Japan has continually been in surplus, with the cumulative total reaching $22 billion in 1981, representing the equivalent of two-thirds of South Korea's total trade deficit. The consistent imbalance has been caused by South Korea's increasing demand for imports of basic and intermediate materials, the result of a policy of export-led high growth based on substantial foreign borrowing. Japan was eminently well placed — historically, linguistically, culturally and of course geographically — to exploit this demand. Also, because of the predominantly 'go-it-alone' character of both countries (indeed, South Korea has much in common in this respect with Meiji Japan's 'rich country, strong army' orientation), very little attempt has been made at integration or a division of labour.

For both countries, the relationship with the other is extremely important, if only for the sake of psychological equilibrium. In view of their outstanding economic prowess, if they can build a truly productive relationship, they could together play a very important role in the region. Although neither of them is rich in natural resources, the similarity in their cultural traditions would make for a good partnership. The former South Korean prime minister, Nam Dok-u, believes that, if the two countries can build a viable division of labour through greater technological cooperation, they could yet conceive a cooperative scheme not unlike that of the EEC.[26]

* * *

The relationship with ASEAN is by and large cordial and productive. Japan has developed institutional arrangements such as the Japan-ASEAN Forum to accommodate the special multilateral character of the relationship. Moreover, the government has been successful in developing domestic consensus on ASEAN's importance for Japan and the need to extend assistance to its member countries; ASEAN, being a multilateral organization, has been helpful in pre-empting the kind of domestic opposition which frequently arises in the case of bilateral relations. Finally, ASEAN's neutralist and self-reliant stand is well adapted to the mood and leanings of Japan.

Japan's over-presence in the region causes difficulties, but there is little that can be done about this. The problem is exacerbated by the trade imbalance which Japan carries with a number of ASEAN countries. The latest attitude survey made in Thailand reveals that the public at large suspects Japan of being out for itself, and that there is a fear of economic domination.[27] Among policy-makers, too, there is a lingering suspicion that Japan is merely out to exploit the region's resources and that, when these are exhausted, it may abandon ASEAN.[28] Some ASEAN countries feel wary of Japan's involvement in the defence of sea-lanes up to 1,000 miles south of Japan, which shows that, four decades after the war, there are still misgivings about Japan's rearmament. However, the fear of the re-emergence of Japan's militarism seems to be on the wane. Thanks to a greatly increased knowledge of Japan and the workings of its political system, few in ASEAN see anything threatening in its current military build-up. Rather, they tend to regard it as a necessary step towards sharing the burden carried by the United States in the region, and towards meeting its own security needs.[29]

On the other hand, ASEAN countries are afraid that Japan may capitulate to the Soviet Union, under either pressure or intimidation, because of its lack of a defence capability. The economic and political implications of a 'Finlandization' of Japan obviously defy prediction. For nations which have so long been used to taking the healthy performance of the Japanese economy for granted, a sudden shift in Japan's orientation would spell serious problems. A Thai political scientist commented that ASEAN wants Japan to rearm to a point, but cannot agree on where that point should be.[30]

It is possible that the people and cultures of Southeast Asia touch a psychological soft spot in Japan. Although there are so many differences, there are also distinct similarities between Japan and the Southeast Asian countries — in their attitude to life, their cultural and religious tolerance, and their unique pragmatism and flexibility. After all, it was ASEAN that prodded Japan away from its hitherto omnidirectional orientation to 'special' and 'heart-to-heart' relations. Since the end of the war, the Japanese have adhered faithfully to their omnidirectional stance, partly because of the global stretch of their economy, but partly because they were afraid of reviving in themselves the pre-war Asianism which was solidly identified with the right-wing militarism of the past. The relationship with ASEAN may yet give an impetus to Japan's further tilt towards Asia, although the fear of alienating its Western and other partners may prevent it from moving too fast or too far in a regionalist direction.

Economism Whither?

In all likelihood, Japan will adhere to its economistic stance for the foreseeable future. Despite pressure from the United States, it will increase its military power sparingly, with caution. The Japanese are conscious that this choice entails continued dependence on the United States, and hence less status for their nation, which is not a comfortable position for a country of Japan's size and influence. However, the results of successive elections seem to show that, barring an unexpected shift in the international scene, the Japanese will prefer to remain economistic rather than go political, let alone military. For all the criticism of Japan being politically passive or reactive, the people seem to approve their government's reluctance to take any action which might precipitate a crisis or instability in the region. They seem to prefer the country to err on the side of inaction.

Japan certainly has more influence today than it had a decade ago, if only because of its economic strength. However, by and large, it chooses to use this influence discreetly. It is known that it has frequently offered advice or suggestions, and has counselled caution to its partners on a variety of questions, including Sino-US relations, the future of Hong Kong and Kampuchean problems, and that the results have been quite effective.[31] Undoubtedly these can be called political initiatives, and they are largely welcomed by the countries in the region. However, Japan prefers to exert its influence behind the scenes, without stating its position publicly. It is a mode of behaviour which seems to suit its current position as a primarily economic power.

Admittedly, the persistent emphasis on economism has taken its toll. For one thing, when such a stance is adopted as the central premise of national policy, it generates a tendency to evaluate every issue on its economic merits. In particular, it blurs the basic purpose of foreign policy issues, and exposes every option to pressure from various internal economic constituencies, whose motives may be largely self-interested. In other words, it makes the country's behaviour materialistic and egocentric. What Japan needs, therefore, is to develop a policy framework which is based upon a concept of a new responsibility for Japan — a responsibility which, as a big economy, it ought to embrace. In this way, it would be able to evaluate its policy options in terms of long-range and even 'moral' goals.

However, it would be difficult for Japan to make such a shift because it would entail a consensus on an overall national goal, in the context of an agreed interpretation of the world, something which the

Japanese are loath to see happen. In fact, the economistic orientation developed out of a deep-seated fear that the country might one day take up another kind of goal or perception — of a kind that once brought disaster to the whole region. Therefore, any overall national objective, even if it sounds harmless, tends to be looked upon with suspicion in case it, too, might have an ugly side.

It is this state of mind that has bred the 'immobilism' of Japan's policy process, a trait that has provoked different reactions from different quarters. For example, the South Koreans, who have been conditioned to believe that everything should be sacrificed for the sake of political freedom, regard Japan's indecisiveness as a cover-up for selfishness. The United States, too, tends to think that Japan is shirking its responsibility as a member of the free world. However, the ASEAN countries seem to find reassurance in an apolitical Japan, because this fits in with their wish to be free from the interference of outside powers, including Japan, as expressed in the ZOPFAN concept. China also seems to realize that an economistic Japan serves its current interests better than a politicized one.

The likely course which Japan will pursue is an economistic road with modifications and fine-tuning. In view of its already considerable economic involvement in the region, a modified and improved economism, if applied cohesively, might alter its relationship with the region greatly. Prime Minister Mahathir of Malaysia commented recently that the role of Japan in the region should be as 'a guide and a teacher; a transferer of technology; a role to make these countries richer and more stable, which would benefit in turn not only the region but also Japan and the world'.[32] If that is what the other countries in the region expect, there would be ample ground for Japan to continue its involvement primarily in an economic context. After all, one can contribute best by doing what one is most experienced at and has an aptitude for. The fact that, in the super-austere 1984 budget, ODA expenditure received an allocation of $2.6 billion, an increase of 9.7 per cent over the previous year (the only other increase being in the allocation to defence — 6.5 per cent over 1983), attests to the government's determination, backed by a political consensus, that the nation should try to fulfil its overseas responsibilities.[33]

There are promising signs in the private sector, too, that the country is nearly ready to make the fine-tuning that is necessary. The overwhelming response to the appeals which were made by various philanthropic organizations for assistance for the Kampuchean refugees, or for medical and educational help for the villages in

Southeast Asia, indicates that people are conscious of the need to take the region to their hearts. The media coverage of the region, and the amount of regional literature that is translated into Japanese, have expanded substantially in the last few years. The increasing number of Japanese who are engaged in relief operations, or who study the languages and cultures of the region, coupled with a steady exchange of professors, students, artists and young people, is beginning to create a new infrastructure in Japan for productive interchange with the region. This may well be a sign of Japan's maturing as a nation.

NOTES

Publication data and translations of titles are given in the Bibliography.

Chapter 2: Cold War in Asia

1. Khrushchev, *Khrushchev Remembers: The Last Testament*, p. 239.
2. Westoby, *Communism Since World War II*, p. 78.
3. Huynh Kim Khanh, 'The Vietnamese Communist Movement Revisited', in Institute of Southeast Asian Studies, *Southeast Asian Affairs 1976*, p. 447.
4. Truman, *Memoirs*, vol. 2, pp. 375-6.
5. Ibid., pp. 416-17.
6. Ibid., p. 450.
7. Hosoya, 'Chosen Senso: Genbaku Toka no Kiki', *Rekishi to Jinbutsu*, December 1982, pp. 168-81.
8. Truman, *Memoirs*, pp. 416-17.
9. Nakajima, *Chugoku*, p. 133.
10. Allison, 'American Foreign Policy and Japan', in Rosovsky, ed., *Discord in the Pacific*, p. 21.
11. Truman, *Memoirs*, p. 442.
12. Shibusawa, *Taiheiyo ni Kakeru Hashi*, p. 432.
13. Okabe, 'The Cold War and China', in *The Origins of the Cold War in Asia*, p. 225.
14. Stockwin, *Japan: Divided Politics in a Growth Economy*, pp. 63-5.
15. Ibid., pp. 64-5.
16. Hosoya, 'Yoshida Shokan to Ei-Bei-Chu no Kozu', *Chuo Koron*, November 1982.
17. Stockwin, *Japan*, p. 70.
18. Ibid., p. 71.
19. Ibid., p. 74.
20. Langdon, *Japan's Foreign Policy*, p. 15.
21. Masuda, '1960-nendai Nichibei Keizai Kankei no Seijisei', *Kokusai Seiji*, 1978, pp. 132-53.
22. Ibid., p. 136.
23. Yano, *Tonan-Ajia Seisaku*, p. 56.
24. Far Eastern Economic Review, *Asia Yearbook 1976*, p. 317.
25. Nakajima, *Chugoku*, p. 136.
26. Ibid., pp. 58-65.
27. Ibid., pp. 86, 94-6.
28. Zhukov, *Soren no Ajia Seisaku*, vol. 2, p. 486.
29. Ibid., p. 487.
30. Ibid.
31. Morrison and Suhrke, *Strategies of Survival*, p. 12.

32. Gordon, *Toward Disengagement in Asia*, pp. 24–6.
33. Dore, 'The Prestige Factor in International Affairs', *International Affairs*, April 1975, p. 203.
34. *US News and World Report*, 26 July 1971.
35. Lee and Sato, *US Policy Toward Japan and Korea*, p. 42.
36. Okonogi, 'Chosen-Hanto o Meguru Kokusai-Seiji', in Mitani, ed., *Chosen-Hanto no Seiji-Keizai Kozo*, p. 137.
37. Lee and Sato, *US Policy Toward Japan and Korea*, p. 43.
38. Tsurutani, *Japanese Policy and East Asian Security*, pp. 84–7; and Larsen and Collins, *Allied Participation in Vietnam*, pp. 161–2.
39. Morrison and Suhrke, *Strategies of Survival*, p. 12.
40. Ibid., pp. 15–16.
41. Ibid., p. 17.
42. Ibid., p. 19.
43. Larsen and Collins, *Allied Participation in Vietnam*, pp. 25–51.
44. As cited in Broinowski, *Understanding ASEAN*, pp. 270–2.

Chapter 3: Asia Goes Multipolar

1. Zhukov, *Soren no Ajia Seisaku*, vol. 2, p. 372.
2. Abu Hanifa, 'Revolusi Memakan Anak Sendiri: Tragedi Amir Sjarifudin', in Taufik, ed., *Manusia dalam Kemelut Sejarah*, p. 205.
3. Onghokham, 'Sukarno: Mitos dan Realitas', in Taufik, ed., *Manusia dalam Kemelut Sejarah*, pp. 45–6.
4. For a detailed survey of Gestapu, see Anderson and McVey, *A Preliminary Analysis of the October 1, 1965, Coup in Indonesia*.
5. Sarasin *et al.*, *ASEAN: Problems and Prospects in a Changing World*, p. 3.
6. 'Indonesia's Steady Course in Foreign Policy', *Asian Wall Street Journal* (hereafter *AWSJ*), 31 October 1983.
7. For more detailed discussion of the origins of ASEAN, see Broinowski, *Understanding ASEAN*; Jorgensen-Dahl, *Regional Organization and Order in South-East Asia*; and Morrison and Suhrke, *Strategies of Survival*, Ch. 6.
8. Sarasin *et al.*, *ASEAN*, p. 5.
9. Broinowski, *Understanding ASEAN*, p. 270.
10. Nishihara, *The Japanese and Sukarno's Indonesia: Tokyo-Jakarta Relations 1951–1966*, pp. 131–6.
11. *Nihon Keizai Shimbun*, 9 August 1967.
12. Chawla, *et al.*, eds., *Southeast Asia Under the New Balance of Power*, pp. 1–4.
13. Ibid., pp. 11–12.
14. *New York Times*, 1 April 1968.
15. Kissinger, 'Central Issues of American Foreign Policy', in Kermit, ed., *Agenda for the Nation*, p. 612.
16. Nixon, 'Asia After Vietnam', *Foreign Affairs*, October 1967.
17. *New York Times*, 26 July 1969.
18. Morrison and Suhrke, *Strategies of Survival*, p. 22.
19. Zhukov, *Soren no Ajia Seisaku*, vol. 2., pp. 497–506.
20. Zhou Enlai, in interview with James Reston, *New York Times*, 10 August 1971.
21. *Quandoi Nan Dhan*, 20 July 1971.
22. *New York Times*, 28 February 1972.
23. Koh, 'North Korea: A Breakthrough in the Quest for Unity', *Asian Survey*, January 1973, pp. 83–7; also Korea, National Unification Board, *A White Paper on*

the South-North Dialogue in Korea, pp. 63-109.
 24. Okonogi, 'Chosen-Hanto o Meguru Kokusai-Seiji', in Mitani, ed., *Chosen-Hanto no Seiji-Keizai Kozo*, p. 139. See also Segal, ed., *The Soviet Union in East Asia*, Ch. 6.
 25. Kim Kyung-won, 'South Korea: the Politics of Détente', in Kim Young C. and Halpern, eds., *The Future of the Korean Peninsula*, pp. 55-71.
 26. Korea, National Unification Board, *White Paper*, pp. 135-8.
 27. Kim Kyung-won, 'South Korea', p. 65.
 28. Morrison and Suhrke, *Strategies of Survival*, pp. 181-3.
 29. Ibid., p. 144.
 30. Broinowski, *Understanding ASEAN*, p. 25.
 31. Leifer, *Indonesia's Foreign Policy*, p. 120.
 32. Morrison and Suhrke, *Strategies of Survival*, p. 274.
 33. Broinowski, *Understanding ASEAN*, p. 295.
 34. Ibid., pp. 28-9.
 35. Somsakdi Xuto, 'ASEAN in a Changing World: A Plea for Further Political Cooperation', in Sarasin *et al.*, eds., *ASEAN: Problems and Prospects in a Changing World*, p. 17.

Chapter 4: Japan in the 1970s

 1. Shiota, *Kasumigaseki ga Furueta Hi*, p. 271.
 2. Lee and Sato, *US Policy Toward Japan and Korea*, p. 57.
 3. Ibid., p. 58.
 4. Kawasaki, *Nitchu Fukko-go no Sekai*, pp. 20-1.
 5. Besshi, 'Nitchu Kokko Seijoka no Seiji Taikei', *Kokusai Seiji*, no. 66, 1980, pp. 4-10.
 6. Nakajima, *Chugoku*, p. 217.
 7. Jain, *China and Japan, 1949-1980*, pp. 82-99, 258-64.
 8. A senior Japanese diplomat, in conversation with the author in London, 1982.
 9. Nixon, *US Foreign Policy for the 1970s*.
 10. Ushiba, 'Gendaishi no Shogen', Television interview, NHK, 26 July 1983.
 11. Ibid.
 12. Ibid.
 13. Destler and Sato, *Coping with US-Japanese Economic Conflicts*, pp. 12-13.
 14. Prachoom and Shibusawa, eds., *Asia in the World Community*, p. 128.
 15. *Japan Times*, 15 November 1972.
 16. *Japan Times*, 23 January 1973.
 17. Shibusawa and Saito, eds., *Tonan-Ajia no Nihon Hihan*, p. 6.
 18. For China's arguments in detail, see James Reston's interview with Zhou Enlai, *New York Times*, 10 August 1971.
 19. *International Herald Tribune*, 10 January 1974.
 20. *Guardian*, 14 January 1974.
 21. *Straits Times*, 16 January 1974.
 22. In conversation with the author in Jakarta, 1981.
 23. *New York Times*, 16 January 1974.
 24. Yanagida, *Okami ga Yattekita Hi*, pp. 36-49, 106-11.
 25. Dore, *Energy Conservation in Japanese Industry*, p. 3.
 26. Yanagida, *Okami ga Yattekita Hi*, pp. 65-78.
 27. Ibid., pp. 75-6.
 28. Ibid., p. 182.
 29. Nagano, *Nihon Gaiko Handobukku*, p. 271.
 30. Yanagida, *Okami ga Yattekita Hi*, p. 192.

31. Ibid., pp. 193-4.
32. *Asahi Shimbun*, 16 December 1973.
33. Dore, *Energy Conservation in Japanese Industry*, p. 4.
34. Ibid., p. 3.
35. Ibid., p. 51.
36. For details of Japan's moves to diversify sources in the mid-1970s, see Akao, ed., *Japan's Resource Diplomacy*.

Chapter 5: A New Paradigm of Relations

1. Broinowski, *Understanding ASEAN*, pp. 275-6.
2. Yano, 'Nihon, ASEAN, Tonan-Ajia', in Shibusawa, ed., *Nihon o Mitsumeru Tonan-Ajia*, pp. 19-20.
3. Umarjadi, former director for the Indonesian ASEAN Secretariat, in conversation with the author, Jakarta, March 1976.
4. Broinowski, *Understanding ASEAN*, p. 281.
5. Lee and Sato, *US Policy Toward Japan and Korea*, p. 102.
6. Ibid., p. 103.
7. Stockwin, *Japan: Divided Politics in a Growth Economy*, p. 174.
8. Mendl, *Issues in Japan's China Policy*, pp. 68-75.
9. Swearingen, *The Soviet Union and Postwar Japan*, pp. 87-9, 122-42.
10. Zhukov, *Soren no Ajia Seisaku*, vol. 2, pp. 598-609.
11. Deng Xiaoping, in conversation with a group of Japanese, Beijing, May 1980.
12. Nakajima, *Chugoku*, pp. 220-4.
13. Richard Halloran, *New York Times*, 21 September 1975.
14. For a very clear exposition of this whole complex issue, see Mendl, *Issues in Japan's China Policy*, p. 79, n. 39.
15. Nishiyama, 'Nihon-ASEAN Forum', *Keizai to Gaiko*, May 1977, pp. 36-7.
16. Yano, 'Nihon, ASEAN, Tonan-Ajia', in Shibusawa, ed., *Nihon o Mitsumeru Tonan-Ajia*, p. 32.
17. Nishiyama, 'Fukuda-sori no Tonan-Ajia Rekiho', *Gaiko Jiho*, October 1977, pp. 8-9.
18. Ambassador Hori Shinsuke in Singapore, *Straits Times*, 9 April 1976.
19. Nishiyama, 'Nihon-ASEAN Forum', p. 37.
20. Broinowski, *Understanding ASEAN*, p. 84.
21. Nishiyama, 'Fukuda-Sori no Tonan-Ajia Rekiho', *Gaiko Jiho*, p. 3.
22. Morrison and Suhrke, *Strategies for Survival*, pp. 226-7.

Chapter 6: Persistent Conflicts and Communism in Asia

1. Far Eastern Economic Review, *Asia Yearbook 1983*, pp. 275-6, and *AWSJ*, 16 February 1983.
2. *Yomiuri Shimbun*, 23 February 1983, and *AWSJ*, 3 March 1983.
3. *Yomiuri Shimbun*, 27 February 1983, and *Nihon Keizai Shimbun*, 17 February 1983.
4. Porter, 'Vietnamese Policy and the Indochinese Crisis', in Elliott, ed., *The Third Indochina Conflict*, pp. 76-82.
5. For a more detailed analysis, see Heder, 'The Kampuchean-Vietnamese Conflict' in Elliott, ed., *The Third Indochina Conflict*, pp. 22-42.
6. Lau Teik Soon, 'The Soviet-Vietnamese Treaty', in Institute of Southeast

Asian Studies, *Southeast Asian Affairs 1980*, pp. 54–65.

7. Tretiak, 'China's Vietnam War and Its Consequences', *China Quarterly*, December 1979, pp. 740–67.

8. Frost, 'Vietnam, ASEAN and the Indochina Refugee Crisis', in Institute of Southeast Asian Studies, *Southeast Asian Affairs 1980*, pp. 353–5.

9. Singapore, Ministry of Foreign Affairs, *Vietnam and the Refugees*, p. 23.

10. Frost, 'Vietnam, ASEAN and the Indochina Refugee Crisis', p. 362.

11. Nagano, *Nihon Gaiko Handobukku*, p. 242.

12. 'Who's Winning the War over Kampuchea?', *AWSJ*, 31 January 1983.

13. Leifer, *Indonesia's Foreign Policy*, p. 165.

14. Deng Xiaoping, in conversation with a group of Japanese visitors, Beijing, May 1980.

15. *AWSJ*, 31 January 1981.

16. Rees, *Crisis and Continuity in South Korea*, pp. 24–5.

17. Mitani, ed., *Chosen-Hanto no Seiji-keizai Kozo*, pp. 55–96.

18. Ibid., p. 83–4.

19. *Nihon Keizai Shimbun*, 25 November 1983.

20. Pye, 'The International Position of Hong Kong', *China Quarterly*, September 1983, p. 457.

21. *Far Eastern Economic Review* (hereafter *FEER*), 20 January 1983, pp. 38–42.

22. *FEER*, 7 July 1983 and 21 July 1983.

23. *South China Morning Post*, 15 November 1983.

24. Pye, 'The International Position of Hong Kong', p. 465.

25. *AWSJ*, 17 February 1983.

26. *FEER*, 20 January 1983, pp. 38–42.

27. Ibid.

28. Ibid.

29. *AWSJ*, 27 January 1983.

30. *FEER*, 20 January 1983, pp. 38–42.

31. *Nihon Keizai Shimbun*, 12 December 1983.

32. *AWSJ*, 27 January 1983.

33. *FEER*, 17–23 December 1982.

34. *Time*, 19 December 1983.

35. *FEER*, 17–23 December 1982.

36. Wedel, 'Current Thai Radical Ideology: The Returnees from the Jungle', *Contemporary Southeast Asia*, June 1982, pp. 1–15.

37. Huynh Kim Khanh, 'The Vietnamese Communist Movement Revisited', in Institute of Southeast Asian Studies, *Southeast Asian Affairs 1976*, p. 460.

38. Kennan, 'The Russian Revolution — Fifty Years After', *Foreign Affairs*, October 1967.

39. Ibid.

40. Zhukov, ed., *Soren no Ajia Seisaku*, vol. 2, pp. 367–401, 531–8.

41. Khien *et al.*, eds., *Indochina and Problems of Security and Stability in Southeast Asia*, p. 8.

42. Yano, *Tonan-Ajia Sekai no Ronri*, pp. 92–6.

43. Kaiser *et al.*, *The European Community: Progress or Decline?*, pp. 25, 29.

44. See related discussion in Bertram, 'Japan's Security through the Perspective of a European'.

Chapter 7: The Regional Economy and Interdependence with Japan

1. Krause and Sekiguchi, eds., *Economic Interaction in the Pacific Basin*, pp. 259–62.

2. *AWSJ*, 17 October 1983.
3. Krause, *US Economic Policy Toward ASEAN*, p. 19.
4. *FEER*, 22-28 October 1982.
5. For details of ASEAN/EEC trade, see Harris and Bridges, *European Interests in ASEAN*, pp. 26-36.
6. *AWSJ*, 30 December 1982.
7. *International Herald Tribune — Special Report*, 31 December 1983-1 January 1984.
8. Yano, *Nanboku-mondai no Seiji-gaku*, pp. 110-12.
9. Narongchai Akrasanee, 'ASEAN Economy 1980 — An Overview', in Institute of Southeast Asian Studies, *Southeast Asian Affairs 1981*, p. 11.
10. Hofheinz and Calder, *The Eastasia Edge*, p. 197.
11. Morris-Suzuki, 'Japan's Relations with Southeast Asia since 1970', in *1981 Proceedings of the British Association for Japanese Studies*, p. 149.
12. Hofheinz and Calder, *The Eastasia Edge*, p. 196.
13. United States, Library of Congress, *An Asian-Pacific Regional Economic Organization: An Exploratory Concept Paper*.
14. Krause, 'World Economic Development and Implications for the Pacific Region'.
15. Japan Centre for International Exchange, *Pacific Community Concept*, pp. 3, 14.
16. Pacific Economic Cooperation Conference, *Report of the Standing Committee*, p. 1.
17. Nishiyama, 'Fukuda-sori no Tonan-Ajia Rekiho', in *Gaiko Jiho*, October 1977, pp. 6-7.
18. *Nihon Keizai Shimbun*, 3 May 1983.
19. *Financial Times*, 15 January 1981.
20. Sekiguchi, *ASEAN-Japan Relations — Investment*, pp. 13-16.
21. Morris-Suzuki, 'Japan's Relations with Southeast Asia', p. 147.
22. Sekiguchi, *ASEAN-Japan Relations — Investment*, pp. 13-16.
23. Robins, 'Japan's Role in Asia: Aid, Trade and Investment Policies', *Phillips and Drew Pacific Survey*, 8 April 1983.

Chapter 8: Japan's Place in the World

1. *Nihon Keizai Shimbun* 25 November 1983.
2. Majonica, 'Bilateral and Multilateral Alliance System of the Two National Sub-systems in Germany and Korea', in Kim Se-jin, ed., *International Peace and Inter-system Relations in Divided Countries*, p. 141.
3. Ibid., p. 142.
4. Deubner, *Industrial Sector and the World Market Today in West German Politics*, p. 37.
5. Ibid., pp. 37-8.
6. *International Herald Tribune*, 25 March 1982.
7. Sinha, *Japan's Role in the 1980s*, pp. 242-7.
8. *Time*, 1 August 1983.
9. Scalapino, 'Perspective on Modern Japanese Foreign Policy', in Scalapino, ed., *The Foreign Policy of Modern Japan*, p. 403.
10. Ushiba, 'A Japanese Perspective on the Trilateral Relationship', in *The State of Trilateral Relations*, pp. 17-18.
11. Bertram, 'Japan's Security through the Perspective of a European', pp. 3-6.
12. Barnett, 'War and Japan: Japan's Concept of Comprehensive National Security', pp. 5-6.

13. Bertram, 'Japan's Security', p. 10.
14. Japan, Prime Minister's Office, *Sogo Anzen Hosho Senryaku*, pp. 79–83.
15. A senior Japanese diplomat, in conversation with the author, London, May 1982.
16. Sato, 'It Takes Two to Tango', *Speaking of Japan*, August 1983, pp. 22–3.
17. Ibid., p. 23.
18. Miyoshi, 'Taiheiyo-Jushi no Senryaku Tenkan', *Chishiki*, Autumn 1983.
19. Deng Xiaoping, in conversation with Japanese visitors, Beijing, May 1980.
20. 'Pax Pacifica', *FEER*, 14 July 1983.
21. In a speech at the World Affairs Council, San Francisco, March 1983.
22. *Nihon Keizai Shimbun*, 30 May, 1 June and 2 June 1983.
23. Japan External Trade Organization, *Chugoku Keizai Shiryo*.
24. For details, see Hilary Conroy, *The Japanese Seizure of Korea, 1868–1910*.
25. Okonogi, 'Japanese Arguments on Korean Politics: A Personal Memorandum'.
26. Statement by Nam Dok-u at the Fifth Korea-Japan Intellectual Exchange Conference, Hakone, Japan, April 1983.
27. Sukhumband, 'Political and Security Dimensions in ASEAN-Japan Relations: Thailand's Perspectives', p. 12.
28. A senior Thai diplomat, in conversation with the author, November 1983.
29. Bruhan, Sukhumband and Zakaria, in their papers and statements at Asia Dialogue Workshop, Bangkok, 23–24 January 1984.
30. Likit Dherawegin, in conversation with the author, in Bangkok, March 1983.
31. A senior Japanese diplomat, in conversation with the author, Kuala Lumpur, March 1983.
32. In an interview with the author, Kuala Lumpur, March 1983.
33. *Nihon Keizai Shimbun*, 25 January 1983.

BIBLIOGRAPHY

Abu Hanifa. 'Revolusi Memakan Anak Sendiri: Tragedi Amir Sjarifudin' (The Revolution Consumes Its Own Child: The Tragedy of Amir Sjarifudin), in Taufik Abdullah, ed., *Manusia Dalam Kemelut Sejarah* (People in the Critical Moments of History), LP3ES, Jakarta, 1978.

Akao Nobutoshi, ed. *Japan's Economic Security*, Gower for the Royal Institute of International Affairs, London, 1983.

Allison, Graham T. 'American Foreign Policy and Japan', in Henry Rosovsky, ed., *Discord in the Pacific: Challenges to the Japanese-American Alliance*, Columbia Books, Washington, 1972.

Anderson, Benedict R., and McVey, Ruth V. *A Preliminary Analysis of the October 1, 1965, Coup in Indonesia*, Cornell University Press, Modern Indonesia Project, New York, 1971.

Barnett, Robert. 'War and Japan: Japan's Concept of Comprehensive National Security', unpublished discussion paper, Washington, 1983.

Bertram, Christoph. 'Japan's Security Through the Perspective of a European', unpublished paper presented at European-Japanese Conference — Hakone VI, Munich, April 1983.

Besshi Yukio. 'Nitchu Kokko Seijoka no Seiji Taikei' (The Politics of Sino-Japanese Normalization), *Kokusai Seiji* (International Politics), no. 66, 1980, pp. 4–10.

Broinowski, Alison, ed. *Understanding ASEAN*, Macmillan, London, 1982.

Bruhan Magenda. 'Political and Security Dimensions in ASEAN-Japan Relations: Perspective from Indonesia', unpublished paper presented at Asia Dialogue Workshop, Bangkok, 23–24 January 1984.

Chapman, J.W.M., Drifte, R., and Gow, I.T.M. *Japan's Quest for Comprehensive Security*, Frances Pinter, London, 1983.

Chawla, Sundershan, Gurtov, Melvin, and Marsot, Alain-Gerard, eds. *Southeast Asia Under the New Balance of Power*, Praeger, New York, 1974.

Conroy, Hilary. *The Japanese Seizure of Korea, 1868-1910*, University of Pennsylvania Press, Philadelphia, 1969.

Destler, I.M., and Sato Hideo. *Coping with US-Japanese Economic Conflicts*, Lexington Books, Lexington, 1982.

Deubner, Christian. *The Industrial Sector and the World Market Today in West German Politics*, Stiftung Wissenschaft Politik, Munich, 1983.

Dore, Ronald. 'The Prestige Factor in International Affairs' in *International Affairs*, April 1975.

—— *Energy Conservation in Japanese Industry*, BIJEPP Energy Paper No. 3, Policy Studies Institute and the Royal Institute of International Affairs, London 1982.

Elliott, David, ed. *The Third Indochina Conflict*, Westview, Boulder, 1981.

Frost, Frank. 'Vietnam, ASEAN and the Indochinese Refugee Crisis', in *Southeast Asian Affairs 1980*, Institute of Southeast Asian Studies, Singapore, 1980.

Gordon, Bernard. *Toward Disengagement in Asia*, Prentice-Hall, New Jersey, 1969.

Harris, Stuart, and Bridges, Brian. *European Interests in ASEAN*, Chatham House Paper No. 19, Routledge & Kegan Paul for the Royal Institute of International

Affairs, London, 1983.
Heder, Steven. 'The Kampuchean-Vietnamese Conflict', in David Elliott, ed., *The Third Indochina Conflict*, Westview, Boulder, 1981.
Hernadi, Andras. *Japan and the Pacific Region*, Trends in World Economy No. 42, Hungarian Scientific Council for World Economy, Budapest, 1982.
Hofheinz, Roy, and Calder, Kent. *The Eastasia Edge*, Basic Books, New York, 1982.
Hosoya Chihiro. 'Chosen Senso: Genbaku Toka no Kiki' (The Korean War: The Danger of Dropping an Atomic Bomb), *Rekishi to Jinbutsu* (History and People), December 1982.
—— 'Yoshida Shokan to Ei-Bei-Chu no Kozu' (The Yoshida Letter and Anglo-American-Chinese Relations), *Chuo Koron*, November 1982.
Huynh Kim Khahn. 'The Vietnamese Communist Movement Revisited', in Institute of Southeast Asian Studies, *Southeast Asian Affairs 1976*, Singapore, 1976.
Institut Français des Relations Internationales. *The State of the World Economy — Rameses 1982*, Macmillan, London, 1982.
Institute of Southeast Asian Studies, *Southeast Asian Affairs 1976*, Singapore, 1976.
—— *Southeast Asian Affairs 1980*, Singapore, 1980.
—— *Southeast Asian Affairs 1981*, Singapore, 1981.
Jain, R.K. *China and Japan, 1949-1980*, Martin Robertson, Oxford, 1981.
Japan, Prime Minister's Office. *Sogo Anzen Hosho Senryaku* (Comprehensive Security Strategy), Tokyo, 1980.
——, Ministry of International Trade and Industry *Keizai Kyoryoku no Genjo to Mondai-ten, 1982* (Current Position and Problems of Economic Cooperation), Tokyo, 1982.
Japan Center for International Exchange. *The Pacific Community Concept*, JCIE Paper, Tokyo, 1982.
Japan External Trade Organization. *Chugoku Keizai Shiryo* (Materials on Chinese Economy), Tokyo, 1983.
Jorgensen-Dahl, Arnfinn. *Regional Organization and Order in South-East Asia*, Macmillan, London, 1982.
Kaiser, Karl, et al. *The European Community: Progress or Decline?*, Royal Institute of International Affairs, London, 1983.
Kawasaki Hideji. *Nitchu Fukko-go no Sekai* (The World After the Sino-Japanese Restoration of Relations), Tokyo, 1972.
Keizai Koho Center. *Japan 1982*, Tokyo, 1982.
—— *Japan 1983*, Tokyo, 1983.
Kennan, George. 'The Russian Revolution — Fifty Years After', *Foreign Affairs*, vol. 46, no. 1, October 1967.
Kermit, Gordon, ed. *Agenda for the Nation*, The Brookings Institution, Washington, 1968.
Khien Theeravit et al., eds. *Indochina and Problems of Security and Stability in Southeast Asia*, Chulalongkorn University Press, Bangkok, 1981.
Khrushchev, N.S. *Khrushchev Remembers: The Last Testament*, translated and edited by Strobe Talbott, Little, Brown, Boston, 1974.
Kim Kyung-won. 'South Korea: The Politics of Détente', in Kim Young C. and Abraham M. Halpern, eds. *The Future of the Korean Peninsula*, Praeger, New York, 1977.
Kim Young C. and Halpern, Abraham M., eds. *The Future of the Korean Peninsula*, Praeger, New York, 1977.
Kissinger, Henry. 'Central Issues of American Foreign Policy', in Gordon Kermit, ed., *Agenda for the Nation*, The Brookings Institution, Washington, 1968.
Koh B.C. 'North Korea: A Breakthrough in the Quest for Unity', *Asian Survey*, January 1973.
Korea, Republic of, National Unification Board. *A White Paper on the South-North Dialogue in Korea*, Seoul, 1982.

Krause, Lawrence B. *US Economic Policy Toward the Association of Southeast Asian Nations: Meeting the Japanese Challenge*, The Brookings Institution, Washington, 1982.
—— 'World Economic Development and Implications for the Pacific Region', unpublished paper presented at the Pacific Economic Cooperation Conference, Bali, November 1983.
Krause, Lawrence B., and Sekiguchi Sueo, eds. *Economic Interaction in the Pacific Basin*, The Brookings Institution, Washington, 1980.
Langdon, Frank C. *Japan's Foreign Policy*, University of British Columbia Press, Vancouver, 1973.
Larsen, Stanley, and Collins, James. *Allied Participation in Vietnam*, US Department of the Army, Washington, 1975.
Lau Teik Soon. 'The Soviet-Vietnamese Treaty', in Institute of Southeast Asian Studies, *Southeast Asian Affairs 1980*, Singapore, 1980.
Lee Chae-jin and Sato Hideo. *US Policy Toward Japan and Korea*, Praeger, New York, 1982.
Leifer, Michael. *Conflict and Regional Order in South-East Asia*, Adelphi Paper No. 162, International Institute for Strategic Studies, London, 1980.
—— *Indonesia's Foreign Policy*, George Allen & Unwin for the Royal Institute of International Affairs, London, 1983.
Majonica, Ernst. 'Bilateral and Multilateral Alliance System of the Two National Sub-systems in Germany and Korea', in Kim Se-jin, ed. *International Peace and Inter-system Relations in Divided Countries*, Research Center for Peace and Unification, Seoul, 1976.
Masuda Hiromu. '1960-nendai Nichibei Keizai Kankei no Seijisei' (Political Aspects of Japanese-US Economic Relations in the 1960s), *Kokusai Seiji* (International Politics), no. 60, 1978.
Mendl, Wolf. *Issues in Japan's China Policy*, Macmillan for the Royal Institute of International Affairs, London, 1978.
Mitani Shizuo, ed. *Chosen-Hanto no Seiji-keizai Kozo* (The Politico-economic Structure of the Korean Peninsula), Kokusai Mondai Kenkyujo, Tokyo, 1983.
Miyoshi Osamu. 'Taiheiyo-Jushi no Senryaku Tenkan' (Strategic Change to Laying Stress on the Pacific), *Chishiki* (Knowledge), Autumn 1983.
Morrison, Charles, and Suhrke, Astri. *Strategies of Survival*, Japanese translation, *Tonan-Ajia Itsutsu no Kuni* (Five Nations in Southeast Asia), Simul, Tokyo, 1981.
Morris-Suzuki, Tessa. 'Japan's Relations with Southeast Asia Since 1970', in *1981 Proceedings of the British Association for Japanese Studies*, BAJS, Sheffield, 1981.
Nagai Yonosuke, and Iriye Akira, eds. *The Origins of the Cold War in Asia*, Columbia University Press/Tokyo University Press, New York and Tokyo, 1977.
Nagano Nobutoshi. *Nihon Gaiko Handobukku* (Japanese Diplomatic Handbook), Simul, Tokyo, 1981.
Nakajima Mineo. *Chugoku* (China), Chuko Shinsho Paperback Series, Tokyo, 1982.
—— *Peking Retsu-Retsu*, 2 vols, Chikuma Shobo, Tokyo, 1981.
Narongchai Akrasanee. 'ASEAN Economy 1980 — An Overview', in Institute of Southeast Asian Studies, *Southeast Asian Affairs 1981*, Singapore, 1981.
Nishihara Masashi. *The Japanese and Sukarno's Indonesia: Tokyo-Jakarta Relations 1951-1966*, University Press of Hawaii, Honolulu, 1976.
Nishiyama Takehiko. 'Nihon-ASEAN Forum' (Japan-ASEAN Forum), *Keizai to Gaiko* (Economy and Diplomacy), May 1977.
—— 'Fukuda-sori no Tonan-Ajia Rekiho' (Prime Minister Fukuda's Southeast Asian Visit), *Gaiko Jiho* (Diplomatic Review), October 1977.
Nixon, Richard. 'Asia after Vietnam', *Foreign Affairs*, October 1976.
—— *US Foreign Policy for the 1970s*, a Report to Congress on 9 February 1972, USGPO, Washington 1972.

Okabe Tatsumi. 'The Cold War and China', in Nagai Yonosuke and Iriye Akira, eds., *The Origins of the Cold War in Asia*, Columbia University Press/Tokyo University Press, New York and Tokyo, 1977.

Okonogi Masao. 'Chosen-Hanto o Meguru Kokusai-Seiji' (International Politics in Korean Peninsula), in Mitani Shizuo, ed., *Chosen-hanto no Seiji Keizai Kozo* (The Politico-economic Structure of the Korean Peninsula), Kokusai Mondai Kenkyujo, Tokyo, 1983.

—— 'Japanese Arguments on Korean Politics: A Personal Memorandum', unpublished paper presented at the Fifth Korea-Japan Intellectual Exchange Conference, Hakone, Japan, 8–10 April 1983.

Onghokham. 'Sukarno: Mitos dan Realitas' (Sukarno: Mystery and Reality), in Taufik Abdullah, ed., *Manusia Dalam Kemelut Sejarah* (People in the Critical Moments of History), LP3ES, Jakarta, 1978.

Pacific Economic Cooperation Conference. *Report of the Standing Committee*, unpublished paper presented at the Pacific Economic Cooperation Conference, Bali, November 1983.

Patrick, Hugh, and Rosovsky, Henry. *Asia's New Giant*, The Brookings Institution, Washington, 1976.

Porter, Gareth. 'Vietnamese Policy and the Indochinese Crisis', in David Elliott, ed., *The Third Indochina Conflict*, Westview, Boulder, 1981.

Prachoom Chomchai and Shibusawa Masahide, eds. *Asia in the World Community*, East-West Seminar, Tokyo, 1973.

Pye, Lucian W. 'The International Position of Hong Kong', *China Quarterly*, September 1983.

Rees, David. *Crisis and Continuity in South Korea*, No. 128, Institute for the Study of Conflict, London, 1981.

Robins, David. 'Japan's Role in Asia: Aid, Trade and Investment Policies', *Phillips and Drew Pacific Survey*, 8 April 1983.

Rosovsky, Henry, ed. *Discord in the Pacific: Challenges to the Japanese-American Alliance*, Columbia Books, Washington, 1972.

Sarasin Viraphol, Amphon Namatra and Shibusawa Masahide, eds. *ASEAN: Problems and Prospects in a Changing World*, Chulalongkorn University Press, Bangkok, 1976.

Sato Hideo. 'It Takes Two to Tango', *Speaking of Japan* (Tokyo), August 1983.

Satoh Yukio. *The Evolution of Japanese Security Policy*, Adelphi Paper No. 178, International Institute for Strategic Studies, London, 1982.

Scalapino, Robert A. 'Perspective on Modern Japanese Foreign Policy', in Robert A. Scalapino, ed., *The Foreign Policy of Modern Japan*, University of California Press, Berkeley, 1977.

Segal, Gerald, ed. *The Soviet Union in East Asia: Predicaments of Power*, Heinemann and Westview for the Royal Institute of International Affairs, London and Boulder, 1983.

Sekiguchi Sueo. *Kantaiheiyo to Nihon no Chokusetsu Toshi* (Japanese Direct Investment in the Pan-Pacific Region), Nihon Keizai Shimbunsha, Tokyo, 1982.

——, ed. *ASEAN-Japan Relations — Investment*, Institute of Southeast Asian Studies, Singapore, 1983.

Shibusawa Masahide. *Taiheiyo ni Kakeru Hashi* (Bridge Across the Pacific), Yomiuri Shimbunsha, Tokyo, 1970.

——, ed. *Nihon o Mitsumeru Tonan-Ajia* (Southeast Asia Looks Towards Japan), Simul, Tokyo, 1977.

Shibusawa Masahide and Saito Shiro, eds. *Tonan-Ajia no Nihon Hihan* (Southeast Asia's Criticism of Japan), Simul, Tokyo 1974.

Shiota Ushio. *Kasumigaseki ga Furueta Hi* (The Days That Shook Kasumigaseki), Simul, Tokyo 1983.

Singapore, Ministry of Foreign Affairs. *Vietnam and the Refugees*, Singapore, July 1979.

Sinha Radha. *Japan's Options for the 1980s*, Croom Helm, London, 1982.
Somsakdi Xuto. 'ASEAN in a Changing World: A Plea for Further Political Cooperation', in Sarasin Viraphol, Amphon Namatra and Shibusawa Masahide, eds., *ASEAN: Problems and Prospects in a Changing World*, Chulalongkorn University Press, Bangkok, 1976.
Stockwin, J.A.A. *Japan: Divided Politics in a Growth Economy*, Weidenfeld & Nicolson, 2nd edn, London, 1982.
Swearingen, Rodger. *The Soviet Union and Postwar Japan*, Hoover Institution Press, Stanford, 1978.
Sukhumband Paribatra. 'Political and Security Dimensions in ASEAN-Japan Relations: Thailand's Perspectives', unpublished paper presented at Asia Dialogue Workshop, Bangkok, 23-24 January 1984.
Taufik Abdullah, ed. *Manusia dalam Kemelut Sejarah* (People in the Critical Moments of History), LP3ES, Jakarta, 1978.
Tretiak, Daniel. 'China's Vietnam War and Its Consequences', *China Quarterly*, December 1979.
Truman, Harry S. *Memoirs by Harry S. Truman*, 2 vols. (vol. 1: '1945: Year of Decision'; vol. 2: '1946-1952: Years of Trial and Hope'), A Signet Book, New York, 1968.
Tsurutani Taketsugu. *Japanese Policy and East Asian Security*, Praeger, New York, 1981.
United States, Library of Congress, Congressional Research Service. *An Asian-Pacific Regional Economic Organization: An Exploratory Concept Paper*, USGPO, Washington, 1979.
Ushiba Nobuhiko. 'Gendaishi no Shogen' (Witness to Contemporary History), Television interview, NHK, 26 July 1983.
—— 'A Japanese Perspective on the Trilateral Relationship', in *The State of Trilateral Relations*, paper presented at the meeting of the Trilateral Commission in Rome, April 1983.
Watanabe Akio. 'Foreign Policy Making: Japanese Style', *International Affairs*, January 1978.
Wedel, Yuangrat. 'Current Thai Radical Ideology: The Returnees from the Jungle', *Contemporary Southeast Asia* (Singapore), vol. 4, no. 1, June 1982.
Westoby, Adam. *Communism Since World War II*, Harvester Press, London, 1981.
Yanagida Kunio. *Okami ga Yattekita Hi* (The Day the Wolf Came), Bunshun-bunko Paperback Series, Tokyo, 1983.
Yano Toru. *Nanboku-mondai no Seiji-gaku* (The Politics of the North-South Problem), Chuko Shinsho Paperpack Series, Tokyo, 1982.
—— 'Nihon, ASEAN, Tonan-Ajia' (Japan, ASEAN, Southeast Asia), in Shibusawa Masahide, ed., *Nihon o Mitsumeru Tonan-Ajia* (Southeast Asia Looks Towards Japan), Simul, Tokyo, 1977.
—— *Tonan-Ajia Seisaku* (Southeast Asian Policies), Simul, Tokyo 1979.
—— *Tonan-Ajia Sekai no Ronri* (The Logics of the Southeast Asian World), Chuko Sosho, Tokyo, 1980.
Zakaria Haji Ahmad. 'Political and Security Dimensions in ASEAN-Japan Relations: A Malaysian Perspective', unpublished paper presented at Asia Dialogue Workshop, Bangkok, 23-24 January 1984.
Zhukov, E.M., ed. *Soren no Ajia Seisaku* (The Soviet Union's Policy in Asia), Japanese translation from Russian original, 2 vols. Simul, Tokyo 1981.

INDEX

Acheson, Dean 15
Adenauer, Chancellor 159
Afghanistan 167, 169
Africa 79, 98, 119
agreements *see also* treaties
   Anglo-Malaysian Defence 57
   China-Japan oil (1977, 1978) 84
   Japan-Vietnam trade (1976) 115
   orderly marketing 72
   preferential tariff 90
   reparations 8, 42, 68
   US-Japan Mutual Defence (1960) 20-1
   US-South Vietnam military assistance 29
   Vietnam peace (1973) 87
agriculture 140, 160
   Ministry of 104
Aichi Kiichi 63
aid 28, 39, 41, 42, 43, 79, 82, 103-5, 109-11, 113-16 *passim*, 119, 122, 154-6, 160, 167, 177
airliner hijackings 120, 161
   shooting down 120
alliances
   Anglo-Japanese (1902) 7
   US-Philippines 60
   US-Thailand 33, 38, 60
anti-hegemony clause 97, 98, 99, 171
Aquino, Benigno 141
Arabs 79-81
arms trade 79, 168
Asian Development Bank 42, 156
Asian and Pacific Council (ASPAC) 45, 87
Association of Southeast Asia (ASA) 38, 39
Association of South-East Asian Nations (ASEAN) 2, 5-6, 34, 37-41, 44-6, 58-61 *passim*, 87-92, 101-10, 112-17 *passim*, 119, 136-44, 146-51 *passim*, 157, 165, 174-5, 177
   Declaration 34, 40-1; of Concord 89, 90, 91
   Industrial Projects 105
   Japan-ASEAN Forum 104
Atkins, Humphrey 123
Attlee, Clement 13
Australia 45, 58, 107, 144, 145, 146

Bailey, Norman 169
balance of power 46-51, 86, 169

Bangladesh 6
bases, military/naval 20, 32, 33, 34, 39, 40-1, 88, 93, 95, 100, 102, 134
Berlin 10
Bhutan 6
Brezhnev, President 91, 97, 98
   Doctrine 50, 86
Britain 6, 7, 13, 19, 25, 39-40, 57, 58, 78, 79, 80, 122, 142
Brunei 6, 84, 142-3, 154
Burma 42, 43, 91, 105, 120

Cambodia *see* Kampuchea
Canada 24, 123, 144
Carter administration 119
Chase Econometrics 142
Chiang Ching-kuo, President 130
China 1-4, 6-8, 11-15 *passim*, 22, 25-7, 32-3 *passim*, 35, 37, 43, 49, 50, 52, 58, 59, 75, 84, 86, 88, 91, 95, 98, 107, 109, 110, 137, 145, 150, 169
   and ASEAN 113
   and Hong Kong 122-8, 134
   and Japan 22, 50, 63-8, 75-6, 86, 95-7, 99, 101, 121, 134, 157, 170-2, 177
   and Korea 13-14, 100, 120-2, 134
   and modernization 2, 6, 66, 109, 122, 157, 171
   and Soviet Union 2, 6, 14, 15, 17, 26, 27, 28, 32, 50, 65, 98, 99, 115, 116, 132, 169; rift 6, 52, 53, 94, 101
   and Taiwan 7, 65-7 *passim*, 124, 127-30
   and US 2, 4, 5, 14-15, 17, 27, 65, 66, 99, 176; rapprochement with 8, 50-4 *passim*, 62, 64, 86, 87, 99, 121, 132
   and Vietnam 27, 51, 110-13, 115, 116
   economy 125-8, 157
Chinese, overseas 7, 113-14
Choi Kyu-hah, President 118
Chun Doo-hwan, President 118-20
CoCom 25
Comecon 91, 112
Cominform 18
Comintern 132
Commonwealth, British 57, 108
communism 3-4, 6, 10-12, 14-15, 19, 26, 28, 35-7, 87, 89, 96, 109-10, 113, 123-5, 128, 130-5
   anti- 16, 23, 29-31, 33, 35-7, 39, 45,

191

87, 91, 103-4, 118, 119, 167
Conferences
 Bandung 98
 Economic Development (MEDSEA) 42
 Geneva refugee 114
 Inter-Parliamentary Union 120
 Pacific Economic Cooperation 147-8
 Youth and Students 11
Czechoslovakia 11, 50

debt 111, 119, 140-1, 146, 155, 170
defence 23, 28, 73, 93-7, 100, 155, 158, 162, 165-8 *passim*, 175, 177
 Five-Power Defence Arrangement 58
 National Defence Programme Outline 95
 Self-Defence Forces 18n., 68, 95
Democratic Socialist Party (DSP) 65
demonstrations, anti-Japanese 8, 73-8, 102, 107; anti-US 20-3, 32
Deng Xiaoping 26, 27, 112-13, 122, 123, 157, 169, 171
detente 46-51, 53, 56, 57, 59
development, economic 5, 6, 22-5, 27, 42, 43, 44, 53, 56, 61, 74, 75, 76, 89, 103, 108, 109, 121, 127-8, 136-7, 139-40, 143-4, 160
Diem, Ngo Dinh *see* Ngo Dinh Diem
Dokto *see* Takeshima
'domino theory' 13, 34, 88, 102
Dore, Ronald 80, 83
Dulles, John Foster 18, 19, 33, 67

East-West confrontation 13, 22, 130, 159
'economism' *see* Japan, economy
economy, regional 5, 73-5, 136-58 *see also individual countries*
EEC 81, 91, 108, 139, 141, 142, 146, 148, 149, 174
Egypt 79, 82
Eisenhower, President 21
energy 78, 83-5, 90, 150, 164, 166
Europe 10-14 *passim*, 56, 117, 133, 145, 161
 East 132-3, 134, 146
 West 1, 8, 10-11, 25, 28, 75, 81, 85, 132, 134, 136, 149, 151, 159-60, 164, 165, 166, 170
Expo '70 62
Faisal, King 78, 82
Far Eastern Commission 17

Fiji 6, 147
Finance, Ministry of 62, 104, 151, 155, 156
Ford, President 92
Foreign Ministry 67-8, 78, 80, 81, 102, 103, 104, 155
France 4, 12, 25, 78, 79, 88
Fukuda, Prime Minister 105, 106, 107, 115, 149, 154, 157
 Doctrine 106-7
Furui Yoshimi 65

Gabon 119
gas, natural 84, 141, 142-3, 146, 150, 154
General Agreement on Tariffs and Trade (GATT) 25, 170
Generalized System of Preferences (GSP) 151
Germany, East 96, 159
 West 1, 7, 8, 10, 22, 28, 78, 120, 159-60, 161, 164
Ghazali Shafie, Tan Sri 60, 61
Gromyko, Andrei 66
Group of 77, 108, 155
Guam Doctrine 48-50, 53, 68, 70, 93

Hagerty, James 21
Heng Samrin 109n., 113, 115
Ho Chi Minh 36
Hong Kong 1, 6-7, 43, 44, 45, 122-8, 129, 134, 136, 137, 138, 144, 145, 148, 150-4 *passim*, 176
Hori Shigeru 98
Hu Yaobang 122, 157
Hua Guofeng 157
Humphrey, Vice-President 31
Hungary 11
Hussein Onn, Datuk 88

ideology 4, 6, 15, 16, 21, 26, 27, 32, 34, 52-3, 56, 68, 75, 97, 104, 107, 109, 133-4, 160, 173
Ikeda Hayato 22-5 *passim*, 32, 41, 42, 160
independence struggles 3, 11, 36, 110, 131-3
India 5, 6, 99
Indochina 1, 6, 87, 88, 91, 103, 106, 108, 110-17 *see also individual countries*
Indonesia 2, 5, 35-40, 42-7 *passim*, 57, 59, 60, 76-8, 84, 88, 90, 98, 108, 115-16, 143
 and Malaysia 36, 42-3, 57
 economy 43, 136-41, 144-6, 148-54

passim
industrialization 2, 90, 105, 127-8, 137, 149, 152-3
International Monetary Fund 41
Inter-Parliamentary Union 53, 120
investment, Chinese 126-7
  Japanese *see* Japan
Iran 82
Iraq 79, 82
Ismail, Tun 58
isolation 7, 8, 85, 158-61
Israel 78-81 *passim*
Italy 7
Ito Masayoshi 167

Japan
  and aid 41, 42, 43, 82, 103-5, 114-16, 119, 154-6, 167, 177
  and ASEAN 45-6, 101-7, 115, 141, 148-50, 157, 165, 174-5; Forum, 104
  and Axis 7
  and Britain 7
  and China 2, 19, 20, 32, 56, 63-8, 96, 97-8, 99, 101, 158, 170-2
  and defence 23, 73, 93-7, 100, 155, 158, 162, 165-8
  and Indonesia 42-3, 76
  and investment 43-4, 145, 146, 148, 151-4
  and Korea 2, 41, 44, 119, 149, 172-4
  and Middle East 78-82
  and oil 78-85, 100, 146
  and region 7, 9, 31, 41-6, 73-8, 99, 100, 101-7, 115, 141, 145-58, 170-5, 177-8
  and Soviet Union 16, 22, 66, 68, 78, 97-8, 99, 100, 101, 158, 166, 175
  and Taiwan 19, 65, 67, 149
  and Thailand 44, 45, 63, 73-6
  and trade 24-5, 33, 42, 44-5, 62, 69-73, 85, 96, 141, 145, 148-51, 162, 164, 171, 174
  and US 7, 8, 16-20, 24-5, 29, 32, 51, 62-4, 66-73, 81, 93-6 *passim*, 100, 101, 102, 158-64 *passim*, 167-8, 176
  and Vietnam 31-3, 56, 68, 115
  demonstrations against 8, 63, 73-8, 102, 104, 107
  economy, 5, 8, 41, 62-3, 71-3, 79, 82-5, 94, 136-8, 148-56, 160, 162; 'economism' 22-5, 41, 69, 71, 79, 160, 176-8
Japan Communist Party (JCP) 17, 95, 101, 131
Japan Socialist Party (JSP) 19-20, 41, 65, 93, 94, 95, 97, 131
Ji Pengfei 124, 125
Johnson, President 30, 46, 47, 49
joint ventures 50, 74, 97, 126, 127, 152-3
Jordan 79

Kampuchea 6, 42, 88, 92, 109, 111-13, 115-17 *passim*, 130, 133, 141, 176
Kaohsiung Export Processing Zone 43
Kasuga Ikko 65
Kawasaki, Hideji 65
Kennedy, President 24, 36
Kenya 119
Khien Samphan 116
Khmer Rouge 6, 112, 113, 116, 133
Khrushchev, Nikita 52, 53
Kim Chong-il 121
Kim Dae-jung 54, 55, 56, 117, 118, 173
Kim Il-sung, President 12, 56, 92, 121, 122
Kim Jong-pil 118
Kim Yong-ju 52, 55, 121
Kim Yong-sam 118
Kishi Nobusuke 20-3 *passim*
Kissinger, Henry 48, 50, 51, 63, 81, 86, 92, 93
Kojima Kiyoshi 146
Komeito 65
Korea 2, 3, 7, 12-15, 51-7, 117-11, 158, 172
  North 1, 5, 6, 8, 12, 14, 15, 31, 51-7, 92-3, 96, 99, 100, 119, 120-2, 173; economy 121
  South 1, 4, 5, 8, 12, 29, 30, 31, 41-5 *passim*, 48, 50, 51-7, 72, 92-3, 100, 107, 117-21, 172-4; and US 4, 16, 31, 53, 92, 93, 119; economy 134, 136, 138, 148-54 *passim*; Yushin Constitution 54, 56, 117
'Koryo, Confederate Republic of' 55
Kosygin, Premier 53
Kukrit Pramoj 88
Kuwait 82

Laird, Melvin 63
Laos 33, 42, 59, 92
Latin America 23, 136, 146
Le Duan 111
Lee Hu-rak 52, 56
Lee Kuan Yew 4, 5, 57, 91
Liberal Democratic Party (LDP) 21, 23, 41, 65, 94, 131

Libya 78
Liu Shaoqi 27
Lomé Convention 149

Macapagal, President Diosdado 43
MacArthur, General 13, 18
MacNaughton, John 29
Mahathir, Prime Minister 57, 177
Malaysia 2, 36-40, 42-6 *passim*, 48, 57-61, 84, 89, 90, 108, 112, 113, 116, 143, 154
  economy 136-8, 140-2, 144-6, 148-52 *passim*
Maldives 6
Malik, Adam 38
Manchuria 7, 8, 158
Manila Pact 108
Mansfield, Mike 163
Mao Zedong 11, 17, 27, 36
Marcos, President 91, 105, 106, 141
Marshall Plan 28; Asian 42
Marx, Karl 131; Marxism 11, 131-2
Matsu 28
Mayaguez incident 93
Meiji era 137, 174
Meyer, Armin 63
Middle East 57, 78-81, 85, 168
Miki Takeo 81, 82, 83, 95, 97, 165, 168
militarism 17, 18, 19, 65, 76, 94, 175
Ministry of International Trade and Industry (MITI) 78, 80, 84, 104, 162
Miyazawa Kiichi 98
Mongolia 68, 134, 166
Mun Se-kwang incident 173

Nakasone Yasuhiro 63, 74-5, 119, 143, 149, 151, 154, 167-8, 170
Nam Dok-u 174
nationalism 3, 18, 77, 78, 88, 131, 133
Nepal 6
Netherlands 79, 81
neutrality 5, 39, 57-61, 91, 93, 94, 95, 102, 107, 147 *see also* ZOPFAN
New Zealand 58, 107, 144
newly industrializing countries (NICs) 44, 47, 129, 136, 138, 142, 149, 154 *see also individual countries*
Ngo Dinh Diem 29-30
Ngo Dinh Nuh 30
Nguyen Duy Trinh 115
Nigeria 119
Nixon, President 5, 23, 48-52 *passim*, 54, 62, 63, 66-71 *passim*, 79, 92, 93, 129
Nolting, Frederick 30

non-aligned movement 5, 91, 107, 108, 162; non-alignment 57, 58, 96, 120
non-tariff barriers 73
'Nordpolitik' 120
North Atlantic Treaty Organization (NATO) 28, 29
North-South relations 147, 164
Northern Territories 16, 78, 97, 99, 101
nuclear policy 168

Ohira Masayoshi 146, 166-7
oil 8, 63, 72, 78-85, 93, 100, 105, 138, 139, 142, 146, 149, 150
Okinawa 24, 32, 64, 66, 68, 70
Olympic games 119, 167
Organization of Arab Petroleum Exporting Countries (OAPEC) 79-82 *passim*
Organization for Economic Cooperation and Development (OECD) 25, 41, 103, 141, 142, 144, 146, 155
Organization of Islamic Countries 108
Organization of Pacific Trade and Development (OPTAD) 146
Organization of Petroleum Exporting Countries (OPEC) 108, 139, 142

Pacific Free Trade Area (PAFTA) 146
Pacific Economic Cooperation Conference (PECC) 147-8
Pacific Trade and Development Conference (PAFTAD) 146
Pak Chung-hee, President 31, 41, 54, 55, 56, 117-18, 173
Pakistan 6, 167
Papua New Guinea 6, 147
peace movement 9, 18, 94-5
'constitution' (1947) 16, 162, 165
Pham Nuo Chuon 133
Pham Van Dong 112, 113
Phan Hien 91, 114
Philippines 2, 38-40, 42-6 *passim*, 60, 88-90 *passim*, 108, 131, 138, 144, economy 136, 137, 140-2, 145, 148-9, 150, 152
Poland 11
polarization 4, 15-22, 31-2, 160
Pol Pot 112, 133
police 17, 18, 20, 21, 23
politics 16-23 *passim*, 25, 36, 46-51, 57-9, 69-73, 75-8, 93-7, 100, 104, 107, 110, 115-18, 120, 125, 129-31, 141, 144
Prapas Charusatien, General 74

Tran Phuong 111
treaties *see also* agreements
   ASEAN Amity and Cooperation (1976) 89, 90, 91
   Britain-China Nanking (1842) 122-3; Peking Conventions (1860, 1898) 122-3
   India-Soviet Friendship (1971) 99
   Japan-Taiwan 65
   Korea-Japan (1965) 41-2, 68
   Nuclear Non-Proliferation 95
   San Francisco Peace (1951) 8, 16, 17n., 18, 19, 20, 68, 158, 161
   Sino-Japanese Peace (1978) 97, 98, 99, 171
   Sino-Soviet Mutual Assistance (1950) 15
   Soviet-Japanese Peace 97, 99
   Soviet-Vietnamese Friendship (1978) 111, 112
   US-Japan Security (1951, 1960) 20-1, 23, 32, 41, 68, 95, 102
trigger price mechanisms 72
trilateralism 164-5; Commission 164
Truman, President 13, 19
   Doctrine 17
Turkey 167
'two camp' theory 18
unemployment 143, 144
United Arab Emirates 82
United Nations 13, 23, 31, 55, 64, 81, 86, 90, 109, 113, 116, 117
   UN Conference on Trade and Development (UNCTAD) 53, 151
United States 2, 4, 7, 12, 33, 35, 36, 48, 58, 79, 86, 87, 92, 99, 123, 132, 141, 142, 144, 145, 146, 149, 151, 172
   and Asia 2, 7, 13, 15, 19, 29-34, 47-9, 62, 68, 70, 71, 87, 93, 163, 172
   and China 2, 3, 13-15, 26, 49, 50, 64, .66, 86, 169, 176; containment of 3, 4, 15, 19, 24, 26, 30, 32, 48, 49, 62, 64; rapprochement with 4, 5, 8, 50-4 *passim*, 62, 64, 86, 87, 99, 121, 132
   and communism 4, 14-15, 19, 23, 29, 30, 32, 64, 167
   and Japan 7, 8, 16-20, 23, 24, 25, 62-4, 67-73, 81, 96, 103, 158-64, 166-70 *passim*, 177
   and Korea 6, 12-15, 31, 53, 72, 92-3, 119
   and Middle East 79, 80, 81

   and Soviet Union 2, 3, 17, 26, 33, 86, 117, 169
   and Taiwan 15, 19, 64, 66, 67, 72
   and Thailand 33-4
   and Vietnam 4, 6, 26, 28-30, 32, 33, 46-7, 86, 87, 88, 116
   demonstrations against 20-3, 32
   International Trade Commission 72

Viet Minh 11
Vietnam 4, 26-34, 89, 91-2, 99, 103, 107, 109n., 110-13, 130, 132-3
   and ASEAN 112; 114-17
   and China 111-12, 113, 115, 116, 130
   and Kampuchea 112-17 *passim*
   and Soviet Union 112, 132-3
   North 3, 5, 6, 11-12, 28, 30, 46, 47, 49, 51, 68, 86, 87, 88, 96, 110
   South 4, 5, 28, 29-32, 42, 47, 86, 87, 88, 110

Wakasugi Kazuo 162
Waldheim, Kurt 114
war
   Arab-Israeli 78-9
   cold 3, 4, 5, 10-34, 52, 54, 66, 68, 94, 96, 106, 117, 133-4, 161, 163, 173
   Indochina 6, 113-15, 130
   Korean 3, 6, 12-15, 18, 33, 53, 117, 173
   Laos civil 33
   'revolutionary' 132
   Russo-Japanese 2, 7
   Vietnam 4, 6, 24, 25, 26-34, 38, 39, 46-7, 53, 57, 69, 75, 86, 96, 99
   World I 7; II 1, 2, 3, 8, 10, 32, 42, 62, 64, 68, 73, 132, 165
West Irian 36
Wharton School or Economics 142
Wijoyo Nitisastro 106
World Health Organization 53
Ye Jianying, Marshal 130
Yoshida Shigeru 18, 19, 42
Yugoslavia 11
   syndrome 14-15

Zengakuren movement 21, 23
Zhao Ziyang 123, 157
Zhou Enlai 50, 51, 53, 65, 67
Zone of Peace, Freedom and Neutrality (ZOPFAN) 57-61, 89-92 *passim*, 94, 177
   Declaration 59-61, 102

Qatar 82
Quandoi Nan Dhan 51
Quemoy 28

Rahman, Tunku Abdul 42, 57
Rajaratnam, S. 114
raw materials 43, 45, 78, 102, 106, 110, 145-6, 148, 149, 150
Razak, Tun 58, 60, 88
Reagan, President 167, 170
rearmament 18-19, 20, 94, 95, 175
refugees 54, 113-15, 177
regional cooperation 5-6, 144-56; South Asian (SARC) 6
reparations 8, 41, 42, 43
Rogers, William 63
Romulo, Carlos 60, 88

Sabah 38, 46, 89, 90, 142
Sanya Dhamasak 76
Sarawak 142
Sasaki Koso 65
Sato Eisaku 32, 41, 45, 63-7 *passim*, 70, 95, 167
Saudi Arabia 78, 79, 80, 82
security 20-3, 46, 91, 93, 95, 98-9, 165-70
 Asian collective 98-9
 'Comprehensive National' 166-7
seikei bunri 25
Senegal 119
Shultz, George 169
Siddhi Savetsila 143
Sihanouk, Prince 59, 116
Singapore 2, 39, 40, 43, 45, 48, 57, 58, 60, 88, 90, 108, 112, 113, 123, 129, 144
 economy 136-40 *passim*, 143, 145, 148, 150, 152-4 *passim*
Snow, Edgar 27
sogo shosha 151, 152
Son Sann 116
South-East Asia Treaty Organization (SEATO) 29, 33, 57, 87
Soviet Union 1, 2, 10-11, 35, 36, 50, 58, 78, 86, 91, 98, 99, 107, 110, 131-5, 159, 165, 166, 168-9
 and Asia 11-12, 91, 98-9
 and China 2, 6, 11, 14, 15, 26, 28, 32, 52, 53, 94, 98, 99, 101, 116
 and Japan 16, 22, 66, 68, 78, 96, 97, 101
 and Korea 12, 14, 100, 120-2
 and US 2, 17, 49, 168-9 *see also* superpower confrontation
 and Vietnam 26, 99, 111, 132-3

navy 98, 134
Spratly Islands 112
Sri Lanka 5, 6
'Stabex' 149
Stans, Maurice 69
strikes 17, 20
Sudan 79
Suharto, President 37, 38, 43, 76, 90
Sukarno, President 35-8 *passim*, 42
summits
 Bali (1976) 89-92, 102, 103
 Group of 77 (1977) 108
 Kuala Lumpur (1977) 105, 107, 149
 Rambouillet (1975) 165
 Tokyo (1964) 43, (1979) 114
 Vienna (1961) 33
 Williamsburg (1983) 170
superpower confrontation 4, 22, 34,-61, 86, 133, 169, 172
 intervention 5, 89
Suslov, Mikhail 26
Suzuki Zenko 167
Syngman Rhee 41

Taiwan 1, 4, 5, 7, 13, 14, 15, 19, 29, 43, 44, 45, 48, 50, 51, 64-7 *passim* 72, 99, 100, 124, 127, 128-30, 144, 158
 and China 7, 65-7 *passim*, 124, 127-30
 and US 4, 16, 51, 129
 economy 136-8 *passim*, 144, 148-52 *passim* 154
Takeiri Yoshikatsu 65
Takeshima 41
Tanaka Kakuei 66, 67, 76, 78, 81, 82, 83, 97, 107
Thailand 2, 3, 4, 29, 33-4, 38-9, 40, 43, 44, 45, 59, 60, 63, 73-6, 77, 78, 88, 89, 91, 110, 112, 114-7 *passim*, 131, 143, 144, 174
 and US 4, 33-4, 38, 39, 49, 88, 108
 economy 136-8, 140-3 *passim*, 145, 148-50, 152-4 *passim*
Thanat Khoman 38, 40
Thanin Kraivichien 106
Thatcher, Margaret 122
Thirayut Boonmee 73
Third World 31, 155, 168, 170
Tonga 147
trade 22, 24-5, 33, 42, 44-5, 62, 69-75, 85, 90, 96, 109, 126-7, 129, 136, 141-3, 145, 148-50, 162-4 *passim*, 170, 171, 174; free trade area 90, 146
'textile wrangle' 69-71
trade unions 17, 20, 23, 97